Doing Interesting Things

With Interesting People

In Interesting Places

# Doing Interesting Things
# With Interesting People
# In Interesting Places

The Building of

u.s. precision lens

*By* Roger L. Howe

# Part One

*The Formative Years*
*Putting the Building Blocks in Place*

# Part Two

*Advancing The Technology*

# Part Three

*Moving Into International Markets*

# Part Four

## *Transitions*

# To

Joyce

Barbara, Karen, Mary and Ed

and

My Associates at U.S. Precision Lens

Who Were Part of The Story

# Foreword

*By* Richard T. Farmer
*Chairman and Founder—Cintas Corporation*

In 1970, Roger L. Howe purchased U.S. Precision Lens, a tiny firm with only twelve employees. It had been in business for forty years. In his own words, this is his story of why he bought it and how he, along with a team of talented associates who joined him, built it into a highly regarded technology company that uniquely serves optic buyers throughout the world. It is told in an engaging, easy to read manner.

Although Roger wrote this book for his associates and family, readers with no connection to him or U.S. Precision Lens will find it fascinating and a pleasure to read. The lessons to be learned from it are significant. Vision, leadership, communication, focus and risk-taking are clear themes that run throughout the pages. The entire history is a "how to" of entrepreneurial endeavor.

To really understand the story that is unfolded here you have to know a lot more about this modest author than he tells about himself. I have known him for forty-five years. We have been friends since our days together at Miami University. Our business careers began about the same time in Cincinnati—he as a sales representative for a Boston paper manufacturer, while I was struggling to make a go of my own small business. Along with our wives, who were classmates and sorority sisters at Miami, we quickly became very close friends. From the very beginning we had many things in common, including an intense interest in business. It was in those early days that I saw qualities in Roger that I knew would lead him to considerable future success. The first indication was in business conversations with people in the group with which we associated. He had an obvious and unique ability to get at the very essence of issues. While others would hear the story, only Roger could put his finger on the crucial points. His communication and marketing skills were remarkable—a clear indication that this man had something special.

When one looks at the influences and experiences that shaped what was to come for him, it is apparent the seeds of success were planted early. First and foremost, he had parents who passed on laudable values and high standards. I knew his exceptional Mother—a wonderful person. Like so many young people who do well in later life, Roger developed a strong work ethic early in life. He was born on a working farm, an environment which quickly gave him adult work responsibilities that caused him to develop lifelong work habits. As he grew he tried to earn money from various farm projects—the kind that are still testing the mettle of today's very best farmers. Some worked, some didn't, and in the process he learned that he could usually analyze the difference before going into ventures that had little hope of success. As a student, in addition to obtaining a good education in college, his entrepreneurial drive surfaced as he and a friend built an extremely successful bicycle sales and rental business.

I was impressed with his pride, hard work and dedication to the paper manufacturer he worked for—S. D. Warren Company. It was clear to me he was destined for a high position with that firm. One day in 1967, Roger called to tell me that Warren had been acquired and would become a division of Scott Paper Company. He was devastated. Scott was a maker of consumer household paper and Warren made the finest printing paper. With such different markets, he felt certain the cultures would be incompatible. From that point on he began to actively consider finding his own business. I advised him against it. We were in our early thirties—he had children—a wonderful home and a great job with a lot of security. His standard of living and work environment was far beyond what he would experience if he left Warren and started a small business. Nevertheless, he was adamant about pursuing his dream.

I saw the situation as a great opportunity to attract this outstanding executive to Cintas, and offered him a significant equity stake to join us. Roger seriously considered it, but in the end came to the conclusion it would not satisfy his entrepreneurial objective.

There is no need for me to say much about his search for a business or what happened after he bought U.S. Precision Lens because it is all laid out in his own words. What is not in the story is specific mention of the considerable personal attributes he brought to bear against the challenges that would be faced in building the company. They are the same attributes seen in all good leaders.

### A Strategic Mind

Roger thinks conceptually. While doing his search he had a clear criteria for the kind of business he was seeking. It had to have the potential to provide unique benefits. This meant it had to be a specialty or niche business far from anything that could be considered a commodity. And, the business had to have a national or even international market because he thought that would make it more interesting. That criteria greatly narrowed the kinds of firms that would be potential acquisitions.

Strategy is evident throughout the story. When looking at USPL he recognized it was viewed as a plastics company that made optics and other things. He quickly repositioned it as an optics company that specialized in making its products with plastic. It was an extremely important mindset which provided the framework for every major decision contributing to the success of the business. It set a new course for what was to follow. For that reason his long time friend and associate, David Hinchman, was exactly right when he stated in a retirement tribute that Roger Howe should really be considered the founder of U.S. Precision Lens.

### A Positive Optimistic Personality

He emphasized what could be accomplished with a confidence in the inherent talent of his co-workers. And, he did it all with the happy advantage of a sunny disposition. There are a lot of different personalities that can and do succeed in business, but some bring ulcers and pressure with them and darken the life for everybody around them. Roger had the enviable ability to inspire terrific results from his team in a constructive environment that had virtually everyone around him eager to help his company succeed.

## A Natural Sales Ability

Imagine what it took to sell lenses made in the United States to companies in Japan—a country that has a highly developed and sophisticated optics industry. At the time Roger and his associates made their first trips to Japan, it was almost laughable that a small company in Ohio could produce lenses that could not be duplicated there for half the price.

Roger has a subtle selling technique, and it seemed to fit the Japanese style. Over time, his potential customers learned to trust him, to respect what he had to say and to believe his company could deliver the unique products it promised. That is how USPL successfully became a valued long-term supplier to some of the most demanding customers in the world.

## Leadership Excellence

Roger had a gift for hiring talented people. He empowered them, focused on the future and let others handle the details. He shared information with the organization, solicited input from everyone and allowed for constructive disagreement. He took risks. He was an innovator who led by example, treating all people with dignity and respect. He freely gave credit to his associates for their contributions. Nothing was more important to him than integrity. His co-workers will tell you, "always do the right thing" was the watchword.

The positive culture at USPL has always been something its people view with pride. It didn't just happen. Nearly twenty-five years ago Roger wrote the *Business Philosophy of U.S. Precision Lens.* This was a mission and values statement produced many years before it was popular to do so. It defined the business, the way the company would conduct itself and how its associates would treat each other. It was the work of a visionary leader that holds true today.

Most of all, he set the direction for the company and consistently led the organization toward the clearly articulated vision of the future.

*     *     *

Now if all this sounds too imposing for the potential entrepreneurs who read it, maybe it shouldn't be. Roger didn't know he had all these skills when he decided to go into business for himself. He probably developed many of them when faced with the harsh necessity to succeed or fail. The purchase of U.S. Precision Lens simply gave him a great opportunity to find out what he could achieve under the most demanding of circumstances.

Roger, and the team that helped him build U.S. Precision Lens, hope that by telling this story others with motivation to create a business will find their experience useful and perhaps even inspiring.

# Preface

In 1987, I delivered a commencement address at Miami University in Oxford, Ohio. The theme reflected my belief that a successful formula for a good life was for individuals to find an activity that, for them, was doing interesting things with interesting people in interesting places. Everyone has a different definition of what satisfies such a quest. For me, it was found in the building of U.S. Precision Lens. That is why I have chosen this title.

Every successful business is built on a fundamentally good idea. Herman Buhlmann's idea for a watch crystal grinding machine led to the founding of the company. It was Henry Buhlmann's good idea to make injection molded plastic optics in the 1950s. Unfortunately, he was never able to find a way to adequately inform the world of the technology's advantages, so only minimal success was achieved for the over fifteen years it languished after incubation.

In 1975, it was our good fortune to invent a way to manufacture unique large aspheric plastic lenses that would be utilized in projection television receivers. It was USPL's third fundamentally good idea and would lead to the company's greatest growth.

We do not have extensive information on the founder and his son who owned the company for thirty-five years. Therefore, the story of the business from 1930 to 1970 is not as detailed as I would like. At the other extreme, it is not possible to name and give credit to all the people who made significant contributions from 1970, when I purchased the business, to the present, but there are hundreds of them. This is unfortunate, and the one aspect of the effort I regret. Many highly dedicated and creative people that were once or are now a part of U.S. Precision Lens did a multitude of things that truly made a difference.

I have chosen to divide the story into three periods:

- 1930–1970: when the company was owned by the Bulhmanns and Robert O. Mayer;
- 1970–1989: when I was the majority shareholder and Chief Executive Officer to late 1986, and Chief Executive Officer to mid-1988;
- 1989 and beyond.

Since there is limited information on the first period, and the period from 1989 on is best written by others, this will primarily be our history from 1970 to 1989.

When USPL began to do significant international business in 1979, we became much more aware of exchange rates and economic factors that could affect the company's future. In each chapter covering a specific year, from that point in time on, general economic data has been included to provide additional context to the times and what was influencing our thinking. For the years 1970 through 1999, this information is also delineated in the appendices.

Throughout these pages I have attempted to interweave background, events, strategic thinking, anecdotes and data in a manner that provides the reader an insight into the creation of U.S. Precision Lens and its culture.

# Part One

*The Formative Years*
*Putting the Building Blocks in Place*

# 1

## A Brief Personal History and Search for a Business

This is a business story, and like all such stories, it is really a people story. It is about what drove the entrepreneur and his associates, how they found and developed an idea, how one event led to the next and the people they met along the way.

As the author and the one considered to be U.S. Precision Lens' modern day entrepreneur, for context purposes this begins with a little of my background and quest to own a business. Its purpose is to add dimension for understanding what follows. The real story of the company starts with Chapter 2.

I was born in 1935 to parents who had recently moved to a farm about forty miles west of Cleveland, Ohio and eight miles south of Lake Erie. It was between Wakeman and Birmingham, small towns with populations of 500 and 250 respectively, near where my father was raised. My mother had first been a school teacher and then a commercial artist in Toledo, Ohio before marrying my father in 1923. My father was a sales executive responsible for Ohio and Michigan markets of the S. D. Warren Company, a Boston-headquartered fine paper manufacturer. He had a considerable impact on Warren's success and served for a number of years on its board of directors.

Both of my parents were avid readers and enjoyed the arts. Honesty, hard work and education were values they, like most parents, imparted to their children. I saw two worlds as I grew up: the farming and blue collar world in the area where we lived, and another with the many successful business people who frequently visited our home, or whom we visited in Cleveland and other cities. My parents had very high personal standards and insisted that everyone be treated with dignity and respect. They particularly abhorred race and religious prejudices, and would not tolerate such talk.

My first experience with business occurred at about age twelve. A neighbor had an unused chicken coop and all the equipment needed to raise baby chickens. They agreed to rent it to me, and I excitedly explained to my father what a great opportunity it was to go into the chicken business. I also told him I needed to borrow some money to acquire all the things needed for the project. Instead of telling me what he really thought, he said he would be happy to lend the money. However, there was one condition. He asked me to figure out what everything was going to cost, how long it would take to grow the chickens, the cost to feed them, how much I was going to sell them for and, finally, how much profit would be earned after all expenses were subtracted from the sale price.

I enthusiastically did a detailed study and built my first business plan as a condition for the loan. When reporting the results I said, "Dad, there is no way to make money in the chicken business." It was a wonderful business lesson.

We had a second home on the property that housed a farmer my parents had hired to operate our place. Shortly after World War II he moved to California, so my father sold the dairy and the milking machine but kept two cows. I thought it would be fun to milk cows and received encouragement from everyone, particularly my two older brothers. The novelty of the experience lasted about two days, then I hated it. Unfortunately, he kept the cows for another three years. Winter, summer, Christmas, my birthday, and every other day, the cows had to be milked. One of the two or three happiest moments of my life was the day we sold the last cow.

I tried another farming project by entering the hog business. This time my father didn't require a business plan, and it was a mistake. I started the endeavor when hog prices were high. When it came time to sell, prices were low. I later learned that's what most farmers did in those days. In spite of many blue ribbons at county fairs, the bottom line was a net loss of $800—when $800 was a lot of money. It was unclear to me how anybody could make money farming, but it was becoming clear I could not.

The Vermilion River ran through our farm and eventually emptied into Lake Erie. Every spring when the rains came, the river would rise with a swift current. Friends and I annually built a raft to see how far we could ride it before it turned over, came apart or tossed us off. Those annual rafts rides are some of my fondest memories. Looking back, I realize it was incredibly dangerous—but great fun.

In 1951, 1952, and 1953, I planted crops on our farm and on a nearby farm owned by my aunt. I grew soybeans, wheat and corn, and this time I made a little money. Nevertheless, it was the final activity needed to end any possible interest I might have had in becoming a farmer; driving a tractor from one end of the field to the other day after day got old quickly.

Baseball was my greatest love, and if the choice had been mine to select any career, that game most certainly would have been it. Unfortunately, while I could hit and throw fairly well, I was one of the slowest runners in the region.

The farm was a marvelous place to grow up, and while the school system was perhaps not as good as many larger ones with greater resources, I have never had regrets about any of it.

Throughout high school, I had a friend who enthusiastically told me he was going to attend Miami University in Oxford, Ohio. I had never heard of it, but his constant talk about it caused me to become interested. At the last minute, he received a basketball scholarship to another university and I went on to enroll at Miami in the fall of 1953.

Miami was wonderful. In those days students were not permitted to have cars, so everyone had to get along in the small town of Oxford where one couldn't distinguish the rich kids from the poor kids. I was a business major and felt I received an excellent education at what many consider to be one of the most beautiful universities in America.

At the start of my sophomore year I met Joyce Anne Lutz, also a sophomore from Ashland, Ohio. She was a fine arts major, very smart, and great fun to be with. We steadily went together our sophomore and junior years and were engaged our senior year. Two weeks after graduation we were married, and have had a great life together which brought us four wonderful children. As will be explained later, Joyce also did something that made this story possible.

At the beginning of our junior year, two of my friends were starting a retail bicycle business in Oxford. I thought it was a wonderful idea and quickly bought a one-third interest. Shortly thereafter Don Boudinot, one of the partners, believed his grades would suffer, so I bought his interest, along with Don Lerner, who had the original idea. Don Lerner and I, with the eventual help of eight employees, built it into a terrific business that still serves its community today. We sold, rented and repaired what were then called English Lightweight Bicycles. By the fall of our senior year we were buying full semi-truck loads from importers and selling them to students, who occasionally had to stand in line to make the purchase. We thought Miami's no-car rule made a lot of sense. Accounting, marketing, taxes, letters of credit and cash flow took on a whole new meaning. We were getting a first-hand taste of the real world, and it was terrific. Don Lerner later would be the best man in our wedding, the USPL attorney, and a long-time board member. Just before graduating in the spring of 1957, we sold the business to a Miami alum who returned to run it. For us, it was a great learning experience.

Near the end of our final term, one of the senior Boston headquarters executives of S. D. Warren, whom I had known for many years as a result of his frequent visits to the farm, encouraged me to consider Warren and the paper business as a career. Only much later did I learn this was quite a break from tradition, since for over thirty-five years there had been an unwritten rule that sons of executives would not be considered for employment. In any event, I was invited to go to Boston for an interview with this 100-year-old maker of fine-coated, book, and specialty papers that had a superb reputation. Compared to other opportunities, it looked good. Hired as a sales apprentice, I reported to work in Westbrook, Maine two weeks after being married. For most of the first year, the job was to study all sections of the largest papermill to gain a thorough understanding of how paper was made. The process started in the woodyard and ended on the shipping dock.

After a six-month army stint in New Jersey and South Carolina, I was sent to a semester of printing school at the Carnegie Institute of Technology in Pittsburgh. The purpose was to learn graphic arts processes so we would better understand how our printing papers were used. Warren's in-depth training was unique in the industry.

After Carnegie, I was assigned to the Cincinnati office in early 1959. It was a very fortuitous move because Larry Boling, the district manager, was an up-and-coming star who would later become the company's president. He was a superb mentor. We were responsible for representing the company with publishers, printers, paper merchants and advertising agencies in Cincinnati, Columbus, Indianapolis, Louisville, Nashville and points in between. I spent a lot of time from 1959 to mid-1968 in my Volkswagen calling on numerous people, many of whom did not particularly want to see me. There was much success, much rejection, and a lot learned about selling.

In 1962 Larry was promoted to the Boston headquarters and, on his recommendation, I was made District Sales Manager. According to associates, at age twenty-seven I was the youngest person ever to hold that position at Warren, and I approached it with enthusiasm as well as a good dose of apprehension. Shortly thereafter my father passed away, at age sixty-two.

Although I enjoyed and was proud of the association with Warren, I had a strong desire to own a business. It was a subject that was never far from my mind. I tried to buy a paper distributor and a pump manufacturing company in 1966 and 1967. Because of my inexperience in such matters, the attempts were not well executed. My friend Jack Roy (who later would serve on the board of USPL) had left IBM and was building a computer services business. My friend from college days, Dick Farmer, was struggling to make a family industrial laundry succeed. Hours and hours were spent talking with them about the exciting things they were doing as well as with others in similar situations, further heightening my desire to have a business. It seemed that my future at Warren was excellent, so there was an opposite tug.

In 1967, something happened which made it a lot easier to consider alternatives to working for a large company; S.D. Warren was sold to Scott Paper Company. It was not very realistic, but I felt like a baseball player must feel who is traded from one team to another. It was naive, but I was offended. Scott made consumer paper items that sold for a few cents or dollars to people shopping in stores. Warren's papers were used by its customers to further manufacture products for others, and transactions were generally in the thousands of dollars. I said at the time that these were two totally different cultures with totally different markets, and it was my bet that the companies would never be comfortable together. As the future would show, I was right.

At the end of 1967, Dick Farmer offered to sell me a major ownership position in his company, along with the job of running sales and marketing. I nearly resigned from Warren to do it. Ultimately, Dick's attractive offer was rejected because it didn't accomplish the real objective of having my own business. I did, however, in relation to my net worth, make a very large investment in his company that has since had a spectacular payoff. I believed in what he was trying to do—it just didn't meet my need.

Not long after my discussions with Dick, I was promoted to the Boston headquarters as the product manager for printing and publishing papers. About eight months later I was promoted again to advertising manager. In the advertising job, much was learned that would be useful later.

By the fall of 1969, I was still unhappy with Scott's ownership of Warren and what it meant for the future. More importantly, I was still thinking a lot about leaving to buy a business.

In November of 1969, I told Larry Boling I was resigning with the objective of acquiring a business. Although we had talked about this before, this time he understood. I told him to set the date at his convenience. He never did, so I departed on February 1, 1970. The unwritten rule at Warren was that if you ever left, you were not allowed to return. In a complete reversal, the long-time chairman told me to give it a try and if it didn't work out to come on back. It was wonderful to hear those words. After thirteen years at Warren, I was off on a new adventure.

I decided to return to Cincinnati because with all the years there, we had developed a broad range of business and social contacts that would be helpful in pursuing the objective. It also seemed there were many more people in Boston trying to do what I was attempting than in Cincinnati. That was probably a correct assessment.

It was in the months leading up to this decision that Joyce did the thing that made it possible. She loved our home in Concord, Massachusetts and the city of Boston. In spite of that she supported me, knowing that this was something I really had to do. She could have vetoed it, but with no grumbling and at considerable personal sacrifice and risk to our family, she moved back to Cincinnati.

My objective was to buy a business. The criteria for the acquisition was clear. In the search, I never wavered from its three components: specialty in nature, national in market, and not a start-up or a save.

In the years with S.D. Warren, I had learned well the difference between specialty and commodity businesses. When it had $100 million in revenue, earnings were $5 million on 70% of its sales in printing papers and $5 million on the 30% of sales in specialty papers. The example I used at the time, to emphasize the importance of finding a specialty business, was I would lose if I tried to make folding cartons better or less expensively than Container Corporation of America. The goal was to acquire a niche business.

The national in market component came about because I thought it would be far more interesting than simply doing business locally. Start-ups are difficult unless the entrepreneur personally brings technology or a customer base from previous activity to the enterprise. I brought neither. Saves are tough, but perhaps not as much so as start-ups. In the end it could be argued that I did waiver from my criteria and, in fact, did a save.

The volume and profit objective was also clear. It was to eventually have a business that would do $2 million to $3 million in sales and earn $200,000 to $300,000 after taxes. In 1970, it looked like a reasonable but long reach.

Going into this venture, I thought I could stay unemployed for up to a year and a half and invest as much as $200,000 in a company before taking on debt. If it failed I would be severely wounded, but would still have enough resources to make the transition back to employment without too severely affecting the lifestyle of my family. Knowing what had been given up at Warren, I clearly understood failure would have carried with it a large amount of personal devastation.

The approach to finding acquisition candidates was to tell friends and acquaintances what I was trying to do and seek suggestions. Insurance people, lawyers, bankers and accountants were also visited for their input. Soon, some leads started coming in. The first company I tried to buy provided price books to paper merchants throughout the country. Every time paper prices changed, all price books had to be updated; and this firm, at the time owned by the

Mead Corporation, used a computer to make the calculations and provide the copy. Someone else entered a higher bid, causing me huge disappointment. Today, anyone with a low-end personal computer could easily do what that company was doing. Technology eventually would eliminate its value.

I looked at companies making automotive body putty, small industrial springs, polyurethane foam kitchen cabinets and other products. Frankly, the search was much more difficult than I anticipated, and was becoming quite discouraging. The second company I seriously tried to buy manufactured fiberglass parts used by many different businesses. It was not in good shape and I was outbid on this too, which brought more discouragement. To keep from going crazy that summer, I did some marketing work for Dick Farmer while continuing my quest.

In September Joyce was talking on the phone with her friend, Polly Bassett, whose husband Ken, an attorney, interrupted with the instructions to tell me he knew of a company for sale. It was called U.S. Precision Lens and Plastics, Inc. I had heard of it earlier in the summer when John Yaeger, a friend at the First National Bank of Cincinnati, mentioned it could be bought—but quickly added it was a terrible little business not worth pursuing. This time, feeling the pressure to investigate everything, I made an appointment to see the owner Robert O. Mayer and his son-in-law, Philip H. Almand, who was vice-president and general manager.

# 2

## United States Watch Crystal Manufacturing Company
## U.S. Precision Lens and Plastics, Inc. 1930–1970

The company's founder, Herman Louis Buhlmann, was born in Switzerland in 1877. He married Julia Williomher and lived in the town of Tavannes. They had six children, with the eldest, Frank "Henry," being their only son. The family later moved to Solothurn, Switzerland. Herman was a machine designer.

Son Henry, who was born in 1898, studied mechanical engineering at the prestigious Swiss Federal Polytechnic Institute and earned a degree in 1921. After completing a military obligation by serving in the Swiss Army Cavalry, he immigrated to the United States in 1922 and joined the General Electric Company. Not long after Henry came to America, Herman and the rest of his family followed. Herman was employed first by the Bulova Watch Company and then the Gruen Watch Company.

Watch styling demanded cases to be made in a wide variety of shapes, each requiring a precisely-fit glass viewing cover or crystal. The accuracy needed to ensure proper fitting with a minimum of breakage and rejects presented considerable difficulties. Edge grinders used by opticians were quite slow and therefore not suitable for mass production.

Herman had an idea for a machine that in a semi-automatic process would accurately edge-grind the glass crystals. He took it to the management of his Cincinnati employer, the Gruen Watch Company, but it was rejected. In 1929, leaving his family in Cincinnati, Herman returned to his native Switzerland and, with the backing of former Swiss business associates, developed the machine. He supervised construction of the first two along with all the accessory equipment required in the process, before returning to the United States with the blueprints and casting patterns that would allow him to make more. He sold a few machines to former employers, then decided the real opportunity was to go into business for himself.

The United States Watch Crystal Manufacturing Company was incorporated in the State of Delaware in 1930 and started operating at 622 Broadway in Cincinnati. The firm immediately secured all business from the Wadsworth Watch Case Company, then went on to become the largest maker of watch crystals in the nation.

The process was to cut an oversized, flat glass shape of the desired crystal. Next, it was heated with a gas flame on a custom-built rotary sagging machine. When the glass was heat-softened, a plunger pressed a die with the desired curvature against it to make the proper shape. This was followed by annealing to relax the stresses that had been induced. The final step was to employ Herman's machine to edge-grind the exact bevel and dimensions so it would precisely fit the case.

At General Electric, Henry had been successfully designing roller bearings for steel mills. He then became chief engineer for the Rollway Bearing Company in Syracuse, New York. During the years 1922 to 1936, he published a number of technical papers that were considered seminal works in the field.

In 1936, Herman suffered a stroke and, as a result, persuaded Henry to move to Cincinnati to assume management of the company. Herman recovered, and it was suspected that he was much happier designing machines than running the business. He remained active until his retirement in 1950.

During the 1930s and 1940s, U.S. Watch Crystal continued extending its market dominance. Nevertheless, it was still a very small company. World War II was a difficult period because the company's experienced male employees went into the service, requiring Henry to train and employ over twenty women. After the war, twenty of the special grinding machines were in operation as business remained very strong.

In the late 1940s and early 1950s, there was a significant decline and restructuring of the watch manufacturing industry in America. Also, the plastic watch crystal was becoming an important feature in lower-cost watches. It was promoted as being unbreakable. In 1953, Henry purchased one of the first small Van Dorn plastic injection molding machines and started making plastic crystals along with the glass crystals. As the U.S. watch industry continued to decline, he gradually started making plastic parts for other applications.

Parenthetically, a brief tutorial is in order. Thermo-plastic molding starts with the raw material in small pellet form. It is put into a hopper on the molding machine and fed into a heated barrel that contains a rotating screw which, with proper heat and pressure, forces the softened material into a mold. It quickly cools in the mold to a solid state. This is followed by the machine opening and thereby pulling the two mold sections apart, allowing the part to be removed. All of the plastic molding mentioned in these pages utilizes this process.

The Bendix Corporation asked Henry if he could make a plastic lens for its dosimeter, a radiation detection device that was important in the Cold War era. They reasoned that a plastic watch crystal looked somewhat like a lens, and if the back side could be filled in with material, it would function as such. Through trial and error he learned to produce an acceptable Bendix lens, which led to a very large order that probably saved the business. From this beginning the company started building its plastic lens customer base, making toy and instrument lenses. In 1958, the Wadsworth and Gruen watch case factories were closed. It was also the year United States Watch Crystal Manufacturing Company amended its certificate of incorporation to change the name to U.S. Precision Lens and Plastics, Inc. A second injection molding machine was purchased to support what Henry perceived was going to be a growing plastic lens business.

The earliest sales records we have for the company start in 1959, and the only data for 1964 are total sales and profit. From 1959 through 1963, the sales of plastic lenses, molds, glass parts, and profits or (losses) were as follows:

## Sales by Category and Profit

| Fiscal Year | Plastic | Molds | Glass | Total | Profit |
|---|---|---|---|---|---|
| 1959 | $ 9,800 | $ 880 | $20,000 | $31,480 | ($ 2,600) |
| 1960 | 29,000 | — | 32,300 | 61,400 | 6,400 |
| 1961 | 33,900 | — | 27,100 | 61,000 | 1,000 |
| 1962 | 65,700 | 3,400 | 18,900 | 88,000 | 17,000 |
| 1963 | 74,600 | 5,700 | 13,100 | 93,400 | 21,200 |
| 1964 | | | | 98,500 | 21,000 |

Herman Buhlmann died in 1963 at the age of eighty-six. In 1964, Henry, himself then sixty-six, wished to retire.

The assets of the company were sold to Cincinnati resident Robert O. Mayer in July of 1964. The Delaware Corporation was dissolved and the new owner incorporated in Ohio. The price of $50,000 included all machinery, equipment, receivables, raw materials and the trade name. The new officers were as follows:

| | |
|---|---|
| President | Robert O. Mayer |
| Vice-President | Robert W. Gwinner |
| Secretary & Treasurer | Kathleen M. Almand |

One of the first orders of business was to enter into an employment contract with Henry Buhlmann. Bob Mayer was a manufacturer's representative and never became very active in the business. Robert Gwinner was his attorney. Kathleen Almand, Mayer's daughter, was married to Philip H. Almand. Phil, a Georgia native, graduated from Florida Southern College and became an officer in the Navy. He met Kathleen in the South and they lived there after being married. It was speculated that Bob bought the company with the objective of having Phil join it so he could get his daughter back to Cincinnati. In any event, Phil joined the firm in 1965.

Henry Buhlmann remained heavily involved, with his primary task being to teach Phil the business. This arrangement lasted for over two years. In February of 1966, Phil became vice-president and general manager, and the decision was made to buy a third injection molding machine. It was a 75-ton New Britain. (Tonnage of molding machines refers to maximum clamping pressure that can be applied to a closed mold.) At the beginning of 1967, the company entered into a lease with Brand Studios for half of its building at 1216 Central Parkway, and moved to the new location one block south of Music Hall.

Henry taught Phil how to read blueprints, do simple lens designs, quote customers, design mold inserts and just about everything else he needed to know to run the business. Bob Mayer remained quite detached from the operation. Charles Herbol, who had been with the company since the late 1930s, supervised the factory. His expertise was in two areas. The first was operating the glass grinding machines now used only to make instrument windows. The second was cutting, lapping and, finally, precisely polishing the curved mold inserts that formed the plastic lens. Molds were bought from outside tooling firms, and one of the two critical technologies the company possessed was the building of lens inserts—the highly polished curved metal surfaces on which the lens was formed. Its other technology was knowing the special techniques needed to program molding machines to make good lenses. Lens molding requires a process that has far more accuracy than almost all other types of plastic molding.

Sales and profits for the fiscal years ending June 30 from 1965 through 1970 were as follows:

## Sales by Category

| Fiscal Year | Plastic | Molds | Glass | Total | Profit |
|---|---|---|---|---|---|
| 1965 | $ 51,100 | $16,900 | $ 6,900 | $ 74,900 | ($ 6,900) |
| 1966 | 92,300 | 17,800 | 10,700 | 120,800 | 10,700 |
| 1967 | 79,400 | 10,900 | 9,600 | 99,900 | ( 11,100) |
| 1968 | 53,900 | 34,200 | 21,100 | 109,200 | ( 1,400) |
| 1969 | 74,900 | 32,500 | 13,800 | 121,200 | 2,300 |
| 1970 | 122,100 | 32,000 | 21,300 | 172,800 | ( 6,467) |

Revenue in 1970 jumped 43%, primarily because commodity custom molding was being done to fill machine time rather than the creation of new sales of plastic lenses. In April, Mayer hired F. Richard Shockley to be President. Shockley had been with Pennsylvania Optical Company. It was hoped, with his optical background, he could attract new plastic lens business. Also, a new 175-ton New Britain molding machine was purchased that was capable of producing significantly larger optics.

It is not completely clear what happened in the Mayer-Almand relationship with Shockley, but it was not good. With the expensive new machine and a lack of good business, the financial condition deteriorated quickly. By the end of summer, Shockley was gone and Mayer decided to sell the company. This was the profit and loss statement for the year that had ended four months earlier.

### Income Statement—Year Ended June 30, 1970

| | |
|---|---|
| Net Sales | $ 172,833 |
| Cost of Sales | 121,243 |
| Gross Profit | 51,590 |
| Sales, General & Administrative Expense | 59,219 |
| Other Income/Expense | 1,162 |
| Net Loss | $ −6,467 |

To the best of our knowledge, in the forty years the company had been in existence its sales and profits never exceeded $172,833 and $21,200 respectively. Under the Buhlmann ownership, it would appear that it was consistently modestly profitable. Under the six years of Mayer ownership, it had cumulative losses of $12,867 and was only profitable in two of those years.

This was the balance sheet:

### Balance Sheet—As of June 30, 1970

| | | | |
|---|---|---|---|
| Cash | $ 899 | Accounts Payable | $ 19,526 |
| Accounts Receivable | 33,775 | Notes Payable | 47,218 |
| Inventories | 7,583 | Accrued Compensation | 2,982 |
| Prepaid Exp. | 3,298 | Other Accruals | 1,662 |
| Total Current Assets | 45,555 | Total Current Liabilities | 71,388 |
| Other Assets | 7,422 | Long-Term Notes Payable | 68,269 |
| Property & Equip. | 116,170 | Capital Stock | 5,000 |
| Less Accumulated Depreciation | −40,747 | Retained Earnings (Deficit) | −16,256 |
| Net Fixed Assets | 75,423 | Total Shareholders Equity | −11,256 |
| Total Assets | $128,400 | Total Liabilities & Equity | $128,400 |

With less than $1,000 in cash, nearly $34,000 in receivables (much of which was questionable), $115,000 in notes payable and a negative net worth over $11,000, this little company was in terrible shape. Bob Mayer had borrowed the money to buy it, and borrowed the money to keep it going by pledging personal assets. For him, the future looked bleak and he wanted out.

While Herman Buhlmann founded the company, it was Henry's decision to buy a molding machine and eventually make plastic lenses. That was the defining moment for what was to follow.

# 3    The Purchase: 1970

Following up on the Ken Bassett lead, I visited Bob Mayer and Phil Almand at the company in early October. What I saw was somewhat shocking. Everything looked very old, except for the expensive new 175-ton molding machine. Most of the equipment and all of the furniture had an antique-like appearance. Beyond that, the place looked disorganized and unclean. One person said the furniture looked like military surplus from various wars the nation had fought in the last century. It was a terrible first impression, and my impulse was to leave. However, after eight months of unemployment, I desperately needed to get back to work. My search had been fruitless to this point, and I was starting to wonder if it had been a huge mistake to leave Warren.

Bob Mayer and Phil Almand were very cordial as they told me about the business. They represented it as marginal—not making or losing money. I had many questions, and most of the answers were not crisp. This raised more questions, particularly with respect to competition. The intriguing surprise was that the company was doing a small amount of business with Honeywell, IBM, RCA, Polaroid, Xerox, and a few other reasonably important companies. I couldn't understand why great firms like these were doing business with this company that seemed to have such a lack of substance. I needed to find the answer to that question.

They provided me with financial statements, old brochures, and some samples as I left to think about it. The next day I returned and spent a long time with Phil, quizzing him about the competition, customers, and his thoughts as to future prospects. Phil tried to be helpful, but too many of the answers were unclear. He said there were five or six competitors, but with further questioning he could not provide much in the way of details. All were larger companies that, at one time or another, had made some plastic lenses. I finally concluded the only real competitors were a company in England called Combined Optical Industries, and a division of the American Optical Company. Combined had a substantial location disadvantage, and the plastic lens business of American Optical had to be a small part of one of its many divisions. Upon hearing what Phil had to say about it, and drawing on my own experience in a large company, I knew that it was probable their plastic optic business was assigned high overhead allocations and did not get a lot of internal support. The conclusion was that the competitive environment was favorable.

Glass lenses, with comparatively expensive raw material and substantial labor content, were made in individual pieces and had to be mounted in some kind of housing with features that provided for attachment to whatever kind of device they were being used in. This meant additional parts and assembly labor. A compelling case for plastic optics was that the lens, mounts, spacers and brackets could be molded in one piece, eliminating other parts and assembly. Furthermore, multiple units could be made with additional mold cavities. It all seemed to make a lot of sense. However, if it was such a great idea, why had this company never sold more than $122,000 of plastic lenses in the seventeen years it had been molding?

It was obvious they had not figured out a good way to market plastic lenses. The industrial directory, *Thomas Register*, was the only place they advertised. With potential customers throughout the country in a wide variety of industries, it was a formidable problem for a company with such limited resources. I knew anyone buying the company would have to find a good answer to this problem if success was to be achieved.

Phil and I reviewed nearly every customer's volume level and order frequency. While he was quite optimistic, the information was not very encouraging. Once more, I went away to pore over the information and sleep on it. By the next day I had concluded there was no good way in advance to tell whether U.S. Precision Lens and Plastics was a good opportunity. I decided the only way to find out was to buy it and run it for at least six months. The product made sense, there had to be a reason some really good companies were buying from it, and a way had to be found to get to the market—if there was a market.

After a brief negotiation, my attorney, Don Lerner, wrote an agreement that in effect resulted in my buying 90% of the company for $54,000. I estimated my downside risk at $40,000. Phil Almand put in $6,000 for a 10% interest. The agreement was reasonably complicated because Bob Mayer had borrowed all the money to buy it and then added more personally-guaranteed debt to keep it going. The money we put in essentially went to pay off debt and relieve Bob from notes that were, at this point, weighing heavily on him. This was partially done by my personally guaranteeing some remaining debt. The deal was signed October 15. I was president and treasurer, Phil Almand was vice-president, and Don Lerner was secretary. The three of us made up the board of directors.

My search had ended. I was now the thirteenth employee and principal owner of a small, sick company that had much about it I didn't understand. That evening, I went home and wrote the following memo:

---

### October 15, 1970
### Objectives and Strategy for Growth

Our objective is to build a substantial and highly profitable company specializing in the manufacture of precision optics and technical plastic parts. By fiscal 1975, our sales level should reach $1,200,000. After tax, profit should be 15% of invested capital.

The strategy for accomplishing these goals will be to build a strong organization that evaluates every project and major expenditure on its ability to meet the investment return objective. Commodity business will only be solicited when profitability can be demonstrated, and with the understanding it will be eliminated when business that better fits the overall growth plan is available.

Budget objectives will be developed between now and May 1, 1971 for:

1. production
2. sales
3. administration
4. capital expenditures.

In May '71 there will be a complete review of the objectives and strategy. Revisions will be made based on the knowledge developed at that time. In addition, a detailed one-year plan will be prepared. A less detailed three-year plan will also be prepared. The business will then be operated in accordance to the details of the one-year plan.

**October 15–December 31, 1970**

Looking back, the first few days of my ownership were incredible. What follows is told without exaggeration or embellishment. I say this because it would be logical for the reader to wonder.

Phil had hired a part-time bookkeeper. My first objective was to gain a better understanding of the financials, particularly anticipated cash receipts. On the morning of day one, I asked this fellow to provide a list of the accounts receivable showing the aging of each. I then went to the back of the plant to meet with Charlie Herbol, the production manager. It was obvious that Charlie was very knowledgeable about glass-grinding and mold-insert polishing, but not plastic molding. We were running a couple of molding machines on a second shift because reject levels were so high that a customer's shipping schedule couldn't be met with one shift. The original number one machine that Henry Buhlmann bought in 1953 was producing only bad parts. Having seen my first injection molding machine just a few days before, I knew there was a lot to learn.

Upon returning to the office to review the receivables, I couldn't find the bookkeeper. Finally, Phil told me he had quit. Asked why, all Phil could say was he had departed right after the receivables report request. This was not a good sign! Eventually we were able to put the information together, but now there was no one to keep the books.

Phil had hired a young man named Bob Campbell to handle scheduling and shipping. This seemed odd because there was not much to schedule or ship, but Bob was pleasant and tried to be helpful.

In the office, there was a basket full of papers that turned out to be shipping memos for products that had been sent to customers as much as a month earlier but not yet billed. Getting all those invoices sent became the top priority. My first direct order was that customers would be promptly billed at the time of shipping. The last thing I did on the first day was spend fifteen minutes with the second-shift foreman, who was just coming to work. The purpose was to meet him and find out a little about the problems he was having on the two jobs running at night.

The next morning, I inquired as to how production was on the second shift. Bob Campbell said it was good, but we had a new problem—the second shift foreman had quit. For the second day in a row, I had to ask, "Why?" Bob related that he had given a somewhat bewildering reason. The foreman said that he met and talked to Howe for awhile and didn't understand why anyone who seemed to be so smart would buy a company like this. That was his reason for quitting. At this point, I was feeling the guy had clearly overestimated my intelligence!

Phil was trying to be as accommodating as he could, but it was obvious he was mainly interested in designing mold inserts and talking with customers. The details of running a business were far from the top of his list, which partially explained why things were so disorganized and the company had not done better in the past. In spite of this, I have always been grateful to Phil Almand for his cooperation and staying with the company because he was the one person who possessed most of the plastic optic knowledge that Henry Buhlmann had passed along.

Relentlessly, I asked questions about everything, often not getting good answers or even answers at all. After quizzing Phil, Charlie or Bob without success, Bob would usually reappear a while later with a solid answer. It was a pattern that would repeat itself in the days ahead. Bob Campbell quickly became a very important contributor in all matters concerning personnel, machines, production and customer orders.

An employee, Margaret Hughes, told Bob that her husband Milford had worked for the company, been fired a few months earlier, and wanted to come back. His expertise was operating molding machines, and she said he could run them a whole lot better than they were being run now. Apparently, Phil had fired him because he had been doing some target practice with a BB gun one evening in the plant. At this point, a really good machine operator was desperately needed. Milford came to the office to interview with me. At the time I was on a long phone call, so he walked back into the plant. The number one machine, after many programming changes, was continuing to make only bad parts. He made a few adjustments and magically, after days of rejects, it was producing good parts. Bob came to my office to tell me what happened. When Milford arrived a moment later, my first words were, "Milford, you really know how to interview for a job!" Although he had very little formal education, Milford was an artist when it came to programming injection molding machines. Whenever we had a tough problem, he was the one called upon to make the fix. He was a foreman and our top molding technician until March of 1979, when he had to retire prematurely due to severe heart problems.

Don Lerner had written the purchase agreement in a way that provided recourse in the event the company was not found to be substantially as represented. Within three days I realized it was not financially marginal as we had been told, but in fact was losing money. I didn't believe Mayer or Almand had tried to deceive me because it was clear they had little idea as to the details of their financial condition. We had a meeting where the facts were explained as to why losses were being generated. I gave them the option of rescinding the deal, or of having Bob Mayer pick up 50% of any operating losses for the first six months of operation. He did not want the company back and readily accepted the proposal. At the end of the period, this new arrangement cost him only a few thousand dollars.

Something had to be done about the disorganized nature of the office and filthy conditions in the factory. For the first two weekends, Phil, Bob and I came in to clean it up. Phil spent his time getting the papers and the files in the office somewhat organized, while Bob and I went to work on the factory. The company's location in the poor, "over-the-Rhine" district (a reference to an area that was once predominantly inhabited by German immigrants) was an ideal place to get help for the task at hand. Bob went to a couple of local bars first thing Saturday morning and asked if anyone wanted to make some money over the weekend. There were ready takers. He brought back a half-dozen men and put them to work. Within an hour, he fired two for drinking from previously undetected wine bottles.

My friend Dick Farmer sent over a very large bag of shop towels and dust mops from his industrial laundry. Most machines were quite greasy, particularly the molding machines that had long-standing hydraulic oil leaks. Bob's fellows wiped them all clean. The floors were a tougher problem. Around the edges of each machine, there was an oil-soaked material that looked a lot like cat litter. I learned it was a product called Oil Dry, and the company had bought bags of it to soak up the leaks. Apparently, the idea of fixing the oil leaks or at least cleaning them up as they occurred had not been considered. Once the Oil Dry was gone, the team from the local bars scrubbed floors. The following weekend, with another neighborhood group, we painted all the walls.

The place looked better, but not great. It still had the appearance of a second-rate antique shop. Both a Cincinnati student travel company called World Academy and the KDI Corporation had recently declared bankruptcy. Each had high quality office furniture being sold at extremely low prices, so we upgraded. For years, we would say USPL was furnished in "early World Academy and KDI." Given what there was to work with, we had done about as much as was possible to make the premises neat and presentable.

There was still the problem of finding a bookkeeper. Phil knew Beverly Simrall, a person from his church that he thought might be able to help. She had previously performed combined secretarial and bookkeeping work for a small company. After she joined us, the paperwork and books were put in first-class condition quickly. She was a competent hard worker who played a very important accounting and administrative role until leaving in 1977 to take a management position with a medical research institute.

I wanted to visit some prospective customers, so I traveled with Phil to Rochester and upstate New York. At Bausch & Lomb, we were given an opportunity to quote on condenser lenses for microscope lighting systems. At the medical instrument maker, Welch-Allyn, we looked at a number of potential applications for plastic lenses. We also visited Xerox, a long-standing but small customer. From my paper industry experience, I knew what first-class sales presentations looked like, and we were not making them. Our literature was old and outdated. The samples we showed were carried in cardboard boxes or plastic bags. Clearly, we needed to give prospects a much better first impression. In spite of that, not long after this initial trip we received mold and production orders from both Bausch & Lomb and Welch-Allyn. The procedure was to charge the customer for a custom-made mold that we held and also charge for the parts made from the mold.

On returning from the New York trip I had Paul Stewart, a superb graphic artist, design a new logotype and a simple plastic optics brochure. Paul had been a long-time friend and had done extensive graphic design work for Warren when I was advertising manager. The fresh, contemporary logo he did in 1970 continues to be used by the company today on all its letterheads, literature and signage.

We needed to get plastic optics samples into the hands of potential customers so they would better understand the attributes of the technology. In an upstairs storage area, there were numerous boxes containing large quantities of overrun lenses. These had accumulated for years, and many had been made for customers who were no longer active buyers. I found a supplier of clear plastic boxes and ordered a few hundred, with the specification that our new logo be printed on the top. The boxes and samples were taken home, where my children dutifully helped fill each with fifteen to twenty different lenses. For nearly two years thereafter, every time we received an inquiry, a sample box was sent along with the information.

To enhance our presentations, I found a jewelry supply company that sold black, leather-like hinged boxes in which jewelers displayed necklaces and other jewels on black velvet. Many of the lenses had a somewhat jewel-like appearance, and it seemed a good selling strategy for us to handle them as if they were such an item. A few months later I had a skilled woodworker make a beautiful, large rosewood box that fit exactly in a briefcase. It was lined with black velvet and held over fifty samples. Our formerly unimpressive presentations took on almost a ceremonial character. Because we showed and handled the lenses as though they were special, customers tended to react to them in the same manner.

By the end of the first month, I felt the old culture of "take your time to respond" remained far too intact. Frequently, things that had been presumed accomplished were not. Everyone was assembled to hear their frustrated new president say that things had to change and change now. Then I put a sign on the wall that read:

---

**Rule One**

Do what you say you are going to do

When you say you are going to do it

The way you say you are going to do it.

---

My impatience was understood and our reaction time soon improved.

We were molding a large truck taillight lens for the K.D. Lamp Company in the 175-ton machine. K.D. had supplied the mold. Phil had aggressively bid the job because it filled a lot of production time. After it had been running for quite a while, Bob Campbell calculated the raw material usage, production rate and labor costs to find that we were losing money on every piece. I instructed Bob to stop the machine and told Phil to tell K.D. Lamp that we had to raise the price or they had to find another supplier. They chose the latter. This was the first experience that gave me an unsettled feeling about our costing and pricing.

Bob and I spent two weekend days at my home reviewing costs and each customer's production records. First, we calculated our direct and indirect costs to establish hourly machine rates. The desired profit amount was then added. Next, we went through every customer's two or three most recent production records to see if we met the objective. The result from this exercise was that we raised nearly all prices, with the combined average being 88%. It was my unpleasant job to tell the customers.

I called each one, told the contact that I had recently bought USPL, and explained it had been in terrible condition with a high likelihood of going out of business. It was further explained that new cash had been injected to make the company sound, but it could not exist without fixing a pricing problem. I told them we had carefully reviewed our experience with their parts and, after pleading for understanding, informed them what the new price was going to be. To varying degrees, in virtually every case the reaction was similar and unbelievably pleasantly surprising. The predominantly common response was that if we would just answer inquiries, return phone calls, ship to their schedules and, (in some cases), improve our quality, they would gladly pay a higher price. A lot of built-up frustration was vented in those calls, but in only one case was pricing the problem. One customer did not accept the change and removed its mold. The experience further convinced me that the company might be doing something quite special.

Bob had become even more helpful, and was excellent in dealing with customers. He and Milford fixed the leaks on the machines and improved the production cycles. Charlie adjusted the molds that needed better tuning. To my total delight, our production efficiency doubled. I pronounced it Howe's First Law: "If you nearly double your prices and then double your production efficiency, you should make a lot of money!" What we really needed now was sales volume.

Sometime between Thanksgiving and Christmas Phil and I went on another sales trip, this time to New York and Boston. He had hired a manufacturer's representative to sell for us in that region, yet we called on many prospects without much success. The representative impressed me as being more a "pitch man" than a solid sales professional. I concluded then that given the specialized nature of our technology, if we were going to be successful, we had to do business directly with our customers.

In Boston, a very interesting visit was made to Polaroid. We were molding small quantities of a lens for their identification cameras and cutting glass window parts used in camera viewfinder systems. Polaroid had superb plastic lens molding capability, which was used only for their own cameras. Their production went into viewfinders and low-end camera photographic objectives—the lenses that transmitted an image to the film.

In a long meeting with optics-buying specialists and engineers, they asked many questions about measuring, molding machines and coatings. One of the questions involved aspherics—a new term for me, but one that would become of paramount importance to our company. It was an extremely educational call and convinced me they were technologically way ahead of us. The good news was that their experts were strong believers in plastic optics and there was much we could do in the future to improve our capability. One of the participants in that meeting was Robert Bacon, a buyer who would contact us in a little over a year with a wonderful opportunity.

As 1970 came to an end, we were halfway through the fiscal year. Although losses were substantial, the factory was running significantly better and relationships with the customers were improving.

What we needed was more sales volume. I had not yet answered the question as to how our small company, with little resources, could tell its story to design engineers and buyers in many different industries throughout the nation in a cost-effective way. It was the problem I thought about the most.

By now, I had also decided that if we were to be successful in the long run, it would eventually be critical to have control of moldmaking—also commonly referred to as tooling. Since the required machine tools were expensive and skilled moldmakers were difficult to hire, there was not much that could be done about the problem at this time. It was, however, a fundamental conclusion about a future requirement of our business.

After two and a half months of experience, I also concluded that we should not make any plastic parts that did not relate to an optical function. This was a modification of the October 15 objectives and strategy statement. Furthermore, we should always refer to ourselves as an optical company. The world was full of plastic companies, but we would be an optical company that specialized in plastic lenses. We would position ourselves as being very different from the typical plastic molder. Again and again I emphasized to our small staff that USPL was an optical company—not a plastic company. It was an important mindset.

# 5     1971

In January, we placed our first advertisement in the trade magazine *Optical Spectra*. It was a quarter page, black-and-white test to see what kind of response could be generated. The effort was not encouraging because we couldn't identify any specific orders that resulted from the ad. There were a number of inquiries for further information, and we responded with our literature and samples. However, the new business that was trickling in mostly came from customers who found us in the *Thomas Register*.

Our poorly-planned, unattractive 3,800 square-foot downtown location had many limitations, the biggest being no room for expansion. After considering a number of options, I purchased two acres of land at 3997 McMann Road in Clermont County, just off Route 125 past the small village of Withamsville. McMann Road was the last street in that easterly direction to have a Cincinnati address. Everything beyond it was an Amelia address. Having the Cincinnati identification was important because virtually no prospects would know where Amelia was in Ohio. I purchased the land personally, not wishing to further contaminate our already fragile balance sheet.

A contractor was hired to build a well-designed, modern 6,300 square foot factory that included six offices. Bob Campbell, now responsible for all production, helped in the planning. The cost estimate, with mechanicals and other leasehold improvements, totaled $92,000. It would have a positive air pressure (for cleanliness) air-conditioned molding room with space for eight machines, twice our present number. Construction started in late summer and was finished by the end of the year. I personally financed the building portion for the same balance sheet reasons that were followed when the land was purchased.

When I arrived at USPL, Mirro Molds and Plastics was our primary mold or tooling vendor. In fact, we had a very large, long overdue account payable to Mirro for a complicated mold that had been delivered to the company months earlier for a customer that defaulted on its purchase agreement. I immediately issued a check covering Mirro's invoice—to the principal owner's utter surprise. This did wonders for our relationship. However, that spring Wayne Garrett, its very able leader, visited to say he and the other owners had sold their interest in the firm to a company in Franklin, Indiana that would probably use all of its capacity. He planned to move to Franklin, which was about 100 miles from Cincinnati. This was a blow because Wayne knew a little about optics and had been an excellent supplier.

I was already fretting about our lack of control over the moldmaking process before I heard this news. The control problem was really a reaction-time problem that came about because of the way the quoting process worked. A customer would ask us to quote a price for a mold and production parts. We then cost the lens insert while having the mold base and construction quoted by the moldmaker. The moldmaker gave us a price and lead-time estimate based on how busy he was at the moment. We then added the numbers and quoted our customer. By the time we got the job, our moldmaker's backlog frequently had changed substantially, necessitating a new lead-time agreement. Furthermore, it was sometimes useful to have the moldmaker talk directly with the customer regarding design details, an interaction not readily doable with an independent supplier.

If we could not directly control moldmaking, it was important to have a very close working relationship with a first-class moldmaker. With Mirro Molds leaving the area, I set out to find such a new vendor. Having heard excellent things about the Scott Brown Corporation, I met with its president, Bill Brown. We developed a good rapport and made an agreement calling for his firm to be our supplier of choice. He would have an opportunity to quote on all of our jobs, and it was anticipated we would place most of them with his company. He further agreed not to make plastic lens molds for anyone else. This arrangement worked well for about a year. Through no fault of his or ours, we continued to have the administrative, quoting and lead-time problems we experienced with Mirro Molds. It all worked about as well as it could work but was far from what was needed.

Our ability to deal with customers on optical technology questions often was lacking. Phil did a good job considering his limited technical training, but we needed someone with more knowledge. The University of Rochester, with its support from Kodak, Bausch & Lomb and Xerox, had the reputation of being the best undergraduate and graduate university specializing in optics. Late in the spring I went to see Brian Thompson, its optics school dean, to tell him about our business and need for help. He recommended a fellow who had previously worked as a mechanical engineer and would graduate in the summer from their one-year Master's program. Ron Byrd joined the company in September and worked hard to help us upgrade our specialty. Unfortunately, Ron had the optical training but not significant work experience in the field. As a result, our fit together was less successful than either of us had hoped. We took on a couple of complicated projects that were not executed well. In August of 1972, Ron and I mutually decided it would be better if he found employment elsewhere. We had learned a great deal, but now still had a need for optical expertise that would not be answered for some time.

As our fiscal year 1971 ended June 30, eight and a half months after my arrival, much progress had been made. Substantial earlier losses had been erased, and on sales of $188,000 we had a $4,000 profit. In the months of April and May we earned $6,400 and $6,600 respectively. With the capital infusions, long-term debt had gone from $68,000 to $21,000, and shareholders equity had gone from a negative $11,000 to a positive $62,000. While I wasn't impressed with any of these numbers, it appeared that we had turned things around and were on an upward trend. Another sign of confidence was in September, when I put $15,000 of new capital in our children's names into the company.

In August, we changed the name from U.S. Precision Lens and Plastics, Inc. to simply U.S. Precision Lens, Inc. It was still very long and, in my opinion, a little presumptuous for such a small enterprise. The name, however, did have some equity that would be lost with a complete change. On the letterhead and literature, just below the new name, we put in a smaller line that said, "Specialists in Plastic Optics." I was delighted to be rid of the implication that it was a plastics business.

# 6

## An Answer to The Marketing Problem

On May 28, 1971, we received a letter from Mr. Milton G. Leonard, an associate editor of *Machine Design* magazine. He informed us they were going to do a technical article on cost-cutting techniques with optical design for the magazine's readership—some 300,000 design engineers. He asked for our input.

The light bulb went off in my head! Why hadn't I thought of a public relations approach to the marketing problem before this? As advertising manager at S.D. Warren, I worked with two agencies. While the bulk of the budget was for magazine advertisements and printed material, a portion was for public relations. The P.R. objective was to have stories of a more technical nature published on our products and the technologies we were employing. I had been very successful in directly placing such stories only eighteen months earlier by personally visiting editors of printing industry trade publications in the New York area.

A friend who was a media expert in a Cincinnati advertising agency told me that *Machine Design* was far and away the finest magazine in its category, with a diverse and high-quality engineering readership. It was clear that Milt Leonard was offering a great opportunity if we could influence him to make plastic optics an important part of his story and mention USPL along with its Cincinnati location.

I called him to say we would be delighted to help in any way we could, and that he would be receiving samples, literature and photos quickly. I wrote a confirming letter that positioned us as specialists in plastic optics. Phil wrote an excellent letter detailing our manufacturing process, raw materials and cost considerations. Both letters and a large box of samples went out the same day we received his inquiry.

The following week, I called Milt to see if he had received our package and to find out how the article was coming along. He said we were the only people to respond so far. We talked about photos that would be useful, and I asked if they ever used cover photos that related to feature articles. He said they occasionally did, and this provided the opportunity to tell him what I had in mind. Our next-door landlord, Brand Photographic Studios, had a superb photographer who took lens pictures for us. My idea was to create a photo that would have a large number of plastic lenses appear to be floating in space with various colors of light bouncing off them. The photographer thought he could do it. Milt was somewhat interested, so I told him we would have it produced and he could decide if it was useful after seeing it. June 4, seven days after first hearing from him, I sent another letter. The first sentence read, "It is not my intention to overpower you with samples, literature, photographs and telephone calls, even if it seems that way." It was accompanied with more photos that we had taken at his request.

After another week, he again told me that we were the only company to respond to his request. I suggested he do the entire article on plastic optics and that he visit us to see first-hand what it was all about. He agreed to the visit, and a few days later flew from Cleveland, where *Machine Design*'s publisher, Penton Publications, was located. I met his plane and used the ride to Central Parkway to tell him about my recently buying the company, our new factory that was being planned, and more as to why we were so excited about the technology. I also tried to pre-condition him for our very unimpressive facility, knowing that after seeing it he might drop the entire idea. Fortunately, he was intrigued with what we were doing and decided to accept the suggestion to make it, exclusively, a plastic optics story. Shortly thereafter, we sent the color transparency of the floating lenses. He and his art director thought it was wonderful and immediately decided to use it for the cover.

The August 5 issue of *Machine Design*, with our spectacular color photo on the cover and the headline "Plastic Optics," was sent to its 300,000 readers. The six-page story used four photos and three charts that we had furnished. It ended with an acknowledgment of appreciation to U.S. Precision Lens in Cincinnati for all its help in the article's preparation.

I was ecstatic! We had a cover story, with credit, on the advantages of plastic optics in the most desirable trade magazine, and—it had cost us nothing.

We had found the answer to the question of how a small company, with a diverse national market and little resources, finds its customers. It was to educate huge numbers of engineers on our plastic optic capability in trade magazines and then let them find us. To get the *Machine Design* plastic optics article exposure through traditional advertising would probably have cost over twenty times our then net worth.

Anticipating a large response to the story, I prepared additional literature that could be sent with our simple brochure. One piece was entitled "A Word About U.S. Precision Lens."

### Nature of Business

U.S. Precision Lens specializes in the manufacture of molded plastic optics. It is a common reaction of glass lens buyers to view our prices as very low. Conversely, those who are accustomed to buying plastic parts from general molders view our prices as high. We would like to emphasize that we are not in either of these businesses. Our plastic optical expertise requires special talent, equipment and manufacturing conditions.

### Financial and Terms

U.S. Precision Lens is a closely held corporation which has adopted a policy of not releasing financial statements. However, if you desire, we will be pleased to provide a list of our suppliers and the name of our bank. We pay all of our invoices within suppliers' regular terms. We take our terms and conditions of sale very seriously and insist on being paid within them.

### Service

Although plastic optics is our principal interest, we understand that when you buy a component from us, it is one of many used to manufacture your product. We appreciate the need for fast response, tight schedules and delivery credibility. We will do everything we can to help you obtain exactly what you want in the most economical and convenient way.

Four one-page pieces, entitled "News and Ideas from U.S. Precision Lens," were prepared covering plastic optic cost savings success stories, partial tooling service utilizing standard diameters in molds we owned, understanding the economics of plastic optics and our precision glass parts capability. Each was meant to be briefly informative about an aspect of our business.

A few thousand full-color reprints of the *Machine Design* story were ordered with the back cover carrying a letter from me and six additional photos of various types of lenses. It essentially served as a brochure, and because *Machine Design* was the publisher, it provided great credibility.

The article brought a large, high-quality response as contrasted to our earlier advertisement, which attracted only a few not-very-good prospects. There was a noticeable pick-up in business activity, and when fiscal 1972 ended eleven months after publication, our sales had nearly doubled. There were a lot of reasons for this success, but the most important was the *Machine Design* story. In the years ahead I continued to contact publishers of leading engineering trade magazines to place stories, with the only different twist being that USPL people were the authors. With almost every one of them, we were featured on the cover by providing a spectacular 8" x 10" color transparency of interestingly arranged plastic lenses.

We had the following cover stories. In June of 1972, *Electro Optical System Design* had an article by Ron Byrd and Phil Almand titled "Molded Plastic Optics Are Coming." The July 3, 1972 issue of the top magazine in its field with over 200,000 readers, *Electronics*, carried an article I wrote entitled "Plastic Optics, Low Cost Components for Opto Electronic Systems." The publisher in a forwarding section said, "We like to think that it takes aggressive editors with finely-tuned intuitions and generous expense accounts to dig up the kind of really great, exclusive articles and stories that attract and hold a prime audience. And that's usually the case. But, there's occasional serendipity in this business too, in which one story leads to another. For example, take our cover story on plastic optical components by Roger L. Howe." He went on to explain how I contacted them, sent samples, and met with editors. In July of 1975, we authored an article in *Optical Spectra* entitled "Plastic Optics: Economic Moment of Truth." In April 1983, *Machine Design* carried another story. It was written by us and entitled "Plastic Wins Role in High Tech Optics."

There were other articles in publications where we did not have the cover photo. When we had it, we believed readership took a huge jump.

# 7    1972

This was a year perhaps more than any other which brought formative events that set the stage for what would follow. In the preceding fourteen and a half months, much was done to improve the company, but the only significant problem to be solved, at least partially, was the marketing dilemma. We had no new major customer relationships that could provide large sales volume, and the control of moldmaking remained a real issue. Both of these concerns would be answered in 1972.

In January we moved into the new factory on McMann Road. It was a bittersweet move because we had about twenty hourly employees who lived in the "over-the-Rhine" area downtown who would not be coming with us due to a lack of transportation. While thanking them for their service I promised everyone a job, with the offer remaining open for one year if they could figure out how to get to the new location. Unfortunately, not one was ever able to take advantage of it.

The new 6,300 square foot building seemed spacious to us. Although quite small, it was nevertheless 66% bigger than the area we had downtown, and was laid out in a far more efficient manner. It had a tasteful, professional look about it. For the first time we were proud, rather than embarrassed, when visited by customers.

Eight of us from downtown now worked at the new factory. After setting up the equipment, the first priority was to hire a workforce. Bob Campbell found that with little industry in the area, there were many fine people who wanted a job close to home; within two or three days, we had everyone we needed. Training started and before long, we were able to return to production. USPL has always been able to attract superb associates near its Clermont County location.

The many small jobs we were selling added up to a volume level that required a new seventy-five-ton molding machine. It was our fifth machine, and first added since the transfer of ownership. It cost $16,000, a lot of money then, but we needed the capacity.

I had paid for the $33,000 worth of leasehold improvements and building mechanicals personally. Rather than burden the balance sheet with more debt, that advance was converted to common stock in February. In September, another $20,000 was invested in our children's names. Our growth had to be fed with capital.

The prospects calling in response to the *Machine Design* article were usually asking similar questions about molds, raw materials and lens manufacturing. I decided that in addition to our P.R. promotional effort, we should publish a book that would be a primer on our business and answer the most common questions. Hardbound books carry an air of authoritativeness, suggesting the people who publish them know what they are talking about. The plan was to make the project self-liquidating by selling it for $5.00. We would promote it at no cost in the new literature sections of various trade magazines, and distribute it to prospects when making sales calls. We started work on the project in April. Milt Leonard, the editor of the first *Machine Design* article, was hired on a moonlighting basis to edit the copy and to find and oversee an artist who would do the charts and diagrams. Also, he generally advised us on things that could be done to make it more interesting.

Fiscal year 1972 ended June 30 with sales of $348,000, up 85%, and a net profit of $31,000, the most ever earned in the company's forty-two-year history. Equity increased to $153,000, up 247% through profits and capital infusions.

I believe the best way to maximize value for all constituencies in a privately-held business is to run it as though it were a public company. This means having a strong Board of Directors to challenge the owner CEO and add dimension to his thinking on a wide range of issues. It also provides discipline by requiring the organization to think about, prepare and articulate its objectives and plans. From late 1972 on, the Board process was an important part of our management system.

In September, John L. Roy joined our board of directors. We had been good friends since 1964. Jack had been a fast-rising executive at IBM before founding a computer services company called United Data Processing, which became one of the largest regional companies in that field. After selling and running it for new owners, he repurchased it a few years later. Jack's business knowledge, experience and excellent judgment contributed to our success in many ways during the years he served USPL.

Along with having a good board, I believe in thorough audits of a business' books and financial control mechanisms. In 1972, we switched from the small local auditor that had served the previous owners to Arthur Young and Company, then one of what was called the "Big Eight Accounting Firms." Arthur Young did a good job of serving us for a few years, but there was a disagreement with an audit manager over what we perceived to be too much emphasis and time on an area that had little importance. We switched to Arthur Andersen and Company, which did an excellent job with our audit and tax work for years. At the time of our disenchantment, I considered the move a big event because we did not believe in changing accounting, banking, legal and insurance professionals unless there was a strong reason.

That fall, realizing that we were creating technology that might be interesting to others, we had Wood, Herron, and Evans, a Cincinnati patent law firm, develop a secrecy and non-compete agreement which would be signed as a condition of employment for all new associates. It was enforceable because we never hired anyone who had previous plastic optics experience, and Ohio had strong intellectual property protection laws. The agreement required non-competition for three years after leaving employment, and non-disclosure of specific USPL trade secrets forever. We didn't object if people left to go to other optical companies as long as they were not making plastic optics. When it was explained as a job security protection measure, all existing associates signed it without a single objection. Over the years, the agreement has been updated from time to time. Happily, it was never challenged, because we would have had to strongly enforce it. That would have been a distasteful event.

Finally, the answers to the two fundamental concerns and the meeting with a person who would have a huge impact on our future, also took place in 1972. Each deserves a chapter of its own.

# 8     The Moldmaking Solution

In March, just when my concern about the moldmaking control problem was reaching a peak, Wayne Garrett came to see me. He said the company in Franklin, Indiana that had acquired Mirro Molds was not doing well and the arrangement was not working. He was looking at his options and sought my opinion.

I could not have been more pleased because this looked like an opportunity to solve our problem as well as his. We had long discussions about what was required to run a good moldmaking operation. Capital for new machines to increase efficiency and steady work flow were key ingredients. Moldmaking required the highest skill level from people who made tooling. The best were real craftsmen, and fiercely independent. Having a good work environment was an important condition needed to attract them.

Wayne was astonished to see how things at USPL had progressed since he last saw us downtown before Mirro's move to Franklin. I told him all about the things we were doing and wanted to do to improve the business, including the need for moldmaking or tooling control. Finally, I told him that a great answer to his problem and ours would be for us to buy the assets of his business. It was also the best answer to the Franklin owner's problems.

Wayne warmed to the idea, and negotiations proceeded during the next month. A major issue for him was what to do about a small home he and his wife lived in. It was really a luxurious cabin with a beautiful pond in a wooded area not far from the town. When it became obvious that this problem had to be solved if there was going to be a deal, I agreed to buy the property personally for what he had paid for it. Once that was settled, the negotiation was concluded.

On May 16, we closed the deal. We had acquired four lathes, seven vertical milling machines, five grinders, three band saws, an electrical discharge machine (EDM), and twenty-four other machine tool and equipment items. Best of all, the operation came with a top craftsman for a leader and three excellent moldmakers. The price was $45,000. To pay for it, I borrowed the money personally and lent it to the company on a subordinated basis.

Only four and a half months after moving into our new factory, we started construction on the first 2,400 square-foot addition because we wanted to move the moldmaking activity to Cincinnati as quickly as possible.

After signing the papers, Wayne told me I had better start selling molds to other plastic companies because USPL would never be able to keep up with their output. As it turned out, at that very time we sold two molds to a prestigious customer and two molds to a customer who would take us into an exciting new industry where we would be nearly a 100% lens provider. There has never been a time at USPL when we did not have enough new sales to keep our moldmakers busy, and most of the time we also had to go to outside moldmakers to satisfy demand.

In August, the Franklin operation was closed and moved into our new addition. From the day of the purchase and particularly after the move to Cincinnati, it became an unbelievably successful asset. Our lead times for molds shortened dramatically, it was easy to obtain quick quotations, and when necessary, it was no problem to put Wayne in communication with customers. When I concluded in late 1970 that control of moldmaking would be a strategic requirement, little did I realize just what a significant marketing and operational weapon it would become.

Charlie Herbol was continuing to polish the mold inserts but the volume was pushing him. He was still using the techniques that Henry Buhlmann had developed years earlier. We needed to improve that aspect of our technology for efficiency and accuracy. Wayne recognized our shortcomings and had some insight into the problem because he was involved with an amateur group that made telescopes prior to going to Franklin. Telescope-making requires lenses and mirrors that have been ground, polished and tested to extremely high levels of accuracy. Being a moldmaker, he appreciated the skills required to produce anything to tight tolerances.

Near the end of September, Wayne came into my office to say he knew a fellow from his telescope-making group who had exceptional optical knowledge and polishing ability. He urged me to meet Richard J. Wessling because he believed Wessling could help us immensely. The suggestion was enthusiastically accepted and a meeting was arranged soon thereafter.

Dick told me about his background. After high school, he continued his education at the Art Academy of Cincinnati, a very fine institution for art training. He did not pursue an art career but instead became a house painter. His real interests were astronomy, optics and telescope-making. It was obvious he was very knowledgeable about optics and optical fabrication techniques. I told him what we were doing, what our needs were, and that we would like to hire him to lead the optical-finishing activity. He said that he was making $10,000 a year as a house painter and would need that level of income if he were to join us. My response was that was a lot of money for us (at that time), but we needed his skills so we would gladly pay it. We both made a great decision that day.

Dick Wessling has always been at the heart of the USPL business because whenever a customer buys a plastic lens, its power, accuracy and finish have been generated by Dick or someone he trained. He developed techniques for figuring and finishing aspheres that as far as we know, were unique. Beyond that, he and his people routinely produced huge numbers of mold inserts under intense delivery pressure conditions. Over the years, Dick has kept up with the technology and continued to find innovative answers to optical challenges. He did two things that will be mentioned later which had a profound effect on the growth of USPL. Wayne Garrett's suggestion to hire Dick Wessling was a fantastic idea!

The moldmaking activity would continue to grow over the years as we poured capital into more and better equipment. Numerically controlled machine tools and electrical discharge machines dramatically improved the manufacturing process. There has never been a doubt in my mind that had we not bought Mirro or created something like it, USPL would be a very small company today.

# 9

## Polaroid

In the spring of 1972, we were contacted by Polaroid senior buyer Bob Bacon to determine if we were interested in quoting on two optical parts for a new camera they were developing. Bob was one of the people I met on the call made there in late 1970. The thought of being able to tell prospects that we supplied optics to Polaroid was very exciting.

In 1943, Dr. Edwin Land's three-year-old daughter wanted to know why she couldn't see a picture he had taken. He thought about it, then went to work on the problem. A few years later Polaroid introduced the first instant camera. It used a black and white film that came in a laminate with a developer pouch on one end of each frame. After the picture was taken, the film was pulled out of the camera through a roller system that broke the pouch and evenly spread the developing emulsion. After one minute the laminate was separated, yielding a photo and a messy, gooey piece of paper coated with excess developer. The photo was quickly coated with a preservative—also a sometimes messy process.

Dr. Land perfected the system, and by the late '60s had introduced color film. Photos were fairly good, but the process was still messy. In 1970, Phil Almand obtained an order to cut glass squares and rectangles used as windows in the viewfinders of their top-of-the-line 400 series cameras. The volume accounted for most of our glass fabrication business from 1970 through 1974. In 1972, we had over $50,000 of Polaroid glass sales. That relationship initially gave us a small presence with them.

It had been rumored that the legendary Dr. Land was creating a new instant photo system that would take excellent pictures with no messy peeling and no need for post coating. Bob Bacon confirmed the rumor and told me the parts they were interested in were for the new technology camera. He emphasized that they were on a very tight schedule.

The camera's lighting system design required a photocell. In front of the photocell was a lens which focused the light on it. In front of the photocell lens was a tinted optical wedge with a gear at its base. The optical wedge was used to increase or decrease the light getting through the photocell lens to the photocell so it would be at a proper level. In front of the wedge was a tinted photocell window that needed to be produced in a number of tint densities. The reason for the various densities was that photocell manufacturing had too much tolerance variance. A window with a given density was chosen to properly calibrate the photocell.

We were asked to quote on the photocell lens and the windows in the various densities. The prints were rushed to us and we immediately reported back with our prices. A couple of Polaroid engineers and a buyer (not Bacon) came to Cincinnati to see if we had the capability they needed. They gave us an order for two molds with impressive forecasts for parts. Our new moldmaking team finished the tooling ahead of an already aggressive schedule. Polaroid was impressed, but had a problem: Dr. Land was continuing to redesign many of the parts in the camera. Both of our parts were redesigned a number of times and in each case, we rebuilt the tooling in record time. They had people in our factory every week for two months. On one occasion, Polaroid engineers arrived late in the afternoon with changed drawings, our moldmakers worked most of the night to make the modifications, we were producing parts in the morning, and they were on a plane to Boston by noon. Bob Bacon was so impressed with our cooperation and response time that he asked if we would like to make the optical wedge. Now we had a third part going through modification after modification as Dr. Land perfected his system.

The Polaroid SX-70 was attracting a lot of interest and press coverage as information about it was methodically being leaked to the press. The official announcement came in late October. *Life* magazine ran an eight-page story with Land on the cover taking a picture with his latest invention. The cover headline was "A Genius and His Magic Camera." All photo, business and news magazines as well as newspapers carried stories about it. Just about everyone knew about this remarkable new camera. Among the things that had to be invented to make it work was a flat disposable battery that came in every film pack. It caused much talk because no one had ever heard of such a battery.

The price of the camera was to be $180, and the film was about $.50 a print. The problem was initial demand, and the need for pipeline filling far exceeded supply. To have an orderly introduction, it was first sold in Florida in November and then expanded to other states as quickly as possible. A friend visiting Florida bought us a camera and eight packs of film. It was great fun to take it on sales calls and snap a prospect's picture who was seeing it for the first time. As he held the picture and watched it develop, we were not above pointing out that we worked closely with Polaroid and made three of the optical parts.

A few months after the introduction, I learned from Bob Bacon that other plastic molders had previously been awarded these parts; each had failed. That accounted for the initial urgency. The timing of our Mirro purchase was perfect. We would never have been able to delight Polaroid without that in-house moldmaking capability.

Our business with them grew nicely in the years ahead. We had parts in a follow-on to the SX-70 called the Pronto. Because of cost and capacity concerns, they had us make lenses that otherwise would have been produced internally.

We have always said that Polaroid was our first really large customer. Probably just as important was the credibility we gained because we were a supplier to such a prestigious company.

# 10 Ellis Betensky

In June of 1972 Dr. Fred Abbott, an optical specialist from 3M Corporation, called to arrange a visit. When he arrived he had with him Ellis Betensky, a young man he said was an optical designer. They were trying to determine if a design they had for a copier lens could be executed in plastic.

Ellis was very curious about plastic optics, learning everything he could about the advantages and limitations. He was excited by the design opportunities afforded by plastic aspheres. We mutually concluded that we could not make the 3M part, but I was gratified by Ellis' enthusiasm. He made the statement that plastic was the only major new development in optics in 100 years: that really attracted my attention. When he left my office for a few minutes, I asked Fred to tell me about him. He said Ellis was probably the best optical designer in the United States and in his opinion, the most innovative. I was impressed and concluded this was someone we should know.

Ellis grew up in Iowa and graduated from the University of Iowa majoring in mathematics. He worked as an optical designer at Bausch & Lomb, then Perkin Elmer. After leaving Perkin Elmer to start his own design firm called Opcon Associates, he quickly found a number of companies interested in his services. A key attraction of his firm was a proprietary computer optical design program that he and his associates developed and continually updated.

Ponder and Best was a California importer that distributed Japanese cameras in America. Usually, after building a Japanese brand to a good business level, the maker took over the distribution, eliminating Ponder and Best. To counter this, they changed the name to Vivitar and started selling products under that brand. Vivitar point-and-shoot cameras were designed and made for them in Japan. Most of the business, however, involved interchangeable 35mm fixed focal length and zoom lenses, which Ellis designed. They acquired patent protection on his extremely innovative work and built a substantial business. Ellis became recognized as the premiere zoom lens designer in the world.

While no significant consultative relationship would evolve soon, I made it a point to stay in close touch with Ellis. *The Handbook of Plastic Optics* had bogged down when it became clear that our optical engineer could not write the design chapter. In our first collaborative effort, Ellis agreed to help us on that project.

Ellis Betensky's name will come up again and again in these pages. Our early acquaintanceship led to his becoming one of our most significant contributors, even though he never worked for USPL directly. He would play a critical role by identifying a key technical candidate, designing unique light-emitting diode and video projection lenses, as well as bringing us into a relationship with Vivitar that made it possible to acquire a vital off-shore manufacturing facility.

# 11

## The Hand-Held Calculator and LED Lenses

In late 1971, a few makers of small, simple calculators started selling their products. Mostly, they were offered in office equipment stores. These devices were made possible because, in the 1960s, the machine's brain, a tiny silicon chip, was developed for the missile and space programs. The 1/6-inch square chips crammed in the calculating power of 10,000 transistors. They could add, subtract, multiply, divide and, in the more exotic versions, perform other functions. Large calculators, using fairly bulky gas discharge readout displays, had been around for a longer period. We had been making filter windows for one of the makers, Monroe Calculator.

The exciting thing about the new machines was they were so portable because of battery power and the small dimensions—hence, the name "hand-held calculator." Just as important as the chips in making this possible was the development of the light-emitting diode, or LED. The LED is a semiconductor device that glows brightly when an electrical current is passed through a bonded substrate of a material called gallium arsenide phosphide. It could be miniaturized in a fairly labor-intensive process, and initially cost a few dollars per digit to make. With minimal six-digit calculators, the early displays alone could be over $25.

I had seen these machines advertised, thought it was a great idea, and hoped to have one someday. The most prominent maker in early 1972 was Bowmar. It was a heavy advertiser, particularly in airline magazines. In late April, we received a request for quotation from Bowmar Canada. It came in an envelope that had been rubber-stamped with the return address. The request itself was on a terrible quality, one-page blue mimeographed sheet, also, there were two-part drawings. It was so unprofessional in appearance that discarding it would not have been unreasonable. However, I wondered if Bowmar Canada might be related to the Bowmar that made calculators, so we called the sender to inquire. Indeed, it was a Canadian subsidiary that was responsible for making the LED displays. This marked the beginning of our wild growth ride in the LED lens business.

Bowmar was looking for cylinder lenses with four attachment pins at the corners to fit over nine-digit LED displays. The objective was to magnify the digits so the extremely expensive, gallium arsenide could be reduced in size because at this point in time, unmagnified digits cost over $2.00 each. The Bowmar calculators were selling for $179 at retail, with the display cost being in the $20 range. They also wanted a non-distorting filter window that would be part of the case. For both items, the production forecasts were huge.

We decided that Phil should go to Ottawa to see if all this was real. He met John J. Magarian, Bowmar Canada's tough, hard-driving president. Jack gave Phil a rough time on lead times and some of his responses to their questions. A somewhat shaken Phil returned and I followed up with a couple of calls. We received the orders in May, coinciding with our purchase of Mirro Molds, which allowed us to shorten the lead times. The lens and window prices were approximately 6 cents and 3.5 cents respectively. Now we had plenty of work for the moldmakers. The molds came in on schedule in July, and Bob Campbell put them into full production. Bowmar was very much for real! Jack Magarian would later become a wonderful friend of USPL, and a much bigger customer at another company.

Near the end of September we received a request for quotation from Texas Instruments for cylinder lenses and filter windows. We knew these had to be for calculators. The request listed purchasing and engineering contacts. I called both to tell them about our specialty and that I would bring the quotes to Dallas on October 5. The engineer said, "Great!" and the purchasing fellow said, "Don't come—just send them." I insisted on coming.

With the rosewood sample box full of jewel-like lenses, I went to T.I. First, I saw Don Duff, the engineer responsible for their high-volume consumer calculator. He was impressed with the samples and delighted with our short lead time quotes. He said, "Let's go see the purchasing guy and get you the orders." The buyer was resistant to moving so fast, and Duff was visibly frustrated with the foot dragging. He insisted that we be given the order and the buyer finally succumbed.

T.I. was also making a twelve-digit engineering calculator that had square root and other scientific features. Don took me to visit Noah Coughlan, its engineer, who was impressed with our capability. He provided drawings and asked for a formal quotation.

This first exhilarating visit to T.I. would be followed by three more in the next two months. We received the twelve-digit engineering calculator lens and window orders. Harry Lee, an engineer responsible for a desktop machine, asked if we could help with a reflection and glare problem on his display window. Remembering from my paper-making days how dull coated papers were manufactured, I had our toolmakers blast the mold surface with a fine sand to impart a uniform surface that was not shiny. It worked beautifully. We called it our proprietary HL finish (for Harry Lee), and sold the feature to everyone.

Looking back, when calendar 1972 ended we had solved the moldmaking problem and acquired three large customers after having none. Our moldmakers were working overtime and we needed more production capacity. Clearly, 1973 would be very exciting.

# 12 1973

As the year began, our one-year old molding room, with space for eight machines that was supposed to take us into the future, now had six—and we were buying two more for delivery as quickly as possible. We were in the first stages of planning a 9,600 square foot addition that would more than double the footage.

In the first week of January I received a call from Mr. John R. Bullock, a prominent senior partner at the law firm of Taft, Stettinius & Hollister. After a gracious introduction, he said he was calling on behalf of a Mr. Sigmund, the Chairman of Combined Optical Industries in Slough, England—a town near London. He said Mr. Sigmund would like to meet me, either in Cincinnati or London, to discuss mutual interests. I knew that Combined was the only other important independent plastic optic maker in the world and probably considerably larger than USPL. It was the only company that we envisioned becoming a significant competitor in our market. I told Mr. Bullock that I would be delighted to meet with him in London and Slough. My objective was simply to learn as much about Combined as possible and benchmark their capability against ours.

My plan was to leave the next week; but having never been out of the country, I needed a passport. In New York one could get a passport in a day; so on January 9th I went there, obtained the passport and spent the afternoon with Ellis Betensky looking at his operation in Stamford. I told Ellis about the trip and asked for his advice as to what should be looked for. After getting his input and a nice dinner, he drove me to the JFK Airport and it was off to London.

Mr. Sigmund met me for dinner at the Churchill Hotel the evening of my arrival. He impressed me as being smart, thoughtful and a very fine gentleman. He was also nearing retirement. His objective for the meeting was exactly the same as mine, except he had an additional goal. He was interested in learning about USPL and me, and it soon became obvious he was looking at us as an acquisition candidate. Not wishing to mislead him, I made it clear that we were still very small and that it was premature for me to have that kind of discussion. Having conveyed that, there was some concern as to whether he would invite me to see their operation the next day, but he did.

After he left that evening, I hailed a cab and told the driver I had never been in London before and would like a tour. Like most London cab drivers, he was very pleasant and knowledgeable. After about an hour of sightseeing my superb guide asked if I would like to visit one of the oldest pubs in the old London section. I readily agreed, and we stopped for one small beer. It couldn't have been a better end to the evening, and I had just made a new friend in one of the great cities of the world.

The next morning Mr. Sigmund drove me to Slough. First we met with his assistant and heir apparent. To my utter surprise, this man was a complete contrast: boastful, brash and arrogant. He said they were going to have an operation in the United States one way or another. Obviously, Mr. Sigmund had briefed him on our lack of interest in being acquired; he, in effect, was making a veiled threat, apparently in hopes of getting me to change my mind.

The Combined factory was probably five times our size. What I saw was quite encouraging. The injection molding machines ranged from small to medium in size, and all looked to be somewhat older. They also used compression molding, which we considered an inferior process. Best of all, they did not have internal moldmaking capability. Like us, before Mirro, they only built the optical mold inserts in-house. It was difficult to assess their size because, in addition to optics, they were molding precision parts and medical items—which indicated some lack of focus on optics.

In time, this trip would become of enormous value because I saw how they made large, thick aspheric magnifiers. While we never used the technology, their technique would later give me an idea as to how we might make large, thick projection video aspheric lenses.

Meeting Sigmund's successor provided a great relief, because I now had no fear of Combined competing with us on our turf.

Mr. Sigmund visited us in May and I extended to him all the courtesies he'd extended to me. In January I told him we had six molding machines, which was the truth. By May we had eight in the molding room, and two that he saw in the glass fabrication room as we passed its open door. He was somewhat shocked as I reluctantly explained that we had new customers requiring all this increased capacity.

Combined Optics never came to the U.S. and was never competition for us. Years later when USPL was far larger, the company was offered for sale to us. However, we declined to even take a look at it.

January 16 was a very important day for us. I was visiting Texas Instruments and invited Noah Coughlan, the engineering leader for scientific calculators, to have dinner that evening. We talked a lot about our respective companies. Finally, I asked him what he would do if he had a company like ours. Without hesitation, he said he would get on a plane and go see all the calculator and LED makers in Silicon Valley. I did exactly what he suggested as quickly as it could be arranged.

Just a few weeks earlier there had been an article in *Business Week* entitled "The Gold That Glitters in Electronic Displays." It was primarily about what was going on with LED's in Silicon Valley— the incredible electronics center south of San Francisco that embraces Palo Alto, Santa Clara, Mountain View, Cupertino and San Jose. It named Fairchild Camera, Hewlett-Packard, Litronix and Monsanto as companies making LEDs and, for three of them, listed their leaders. I called them all, told them about our specialty, told them we were supplying Bowmar and Texas Instruments and that I would like to visit soon. Everyone was interested.

My first call was to Litronix, a very exciting startup led by a thirty-two year old former Monsanto employee. There were five founders, all in their twenties or early thirties, working day and night to make it succeed. It had 2,000 employees, mostly low-cost assemblers in the Far East, and in a very short time had generated $15 million in sales volume. They were selling LED displays to calculator makers around the world and planning to introduce a line of their own. Engineers came from all over the building to see the contents of the rosewood sample box. Blueprints for two lens arrays were given to me for quotation. They called the arrays "bubble lenses" because over each LED digit there was a spherical lens that looked somewhat like a bubble. I phoned Bob Campbell and Wayne Garrett to describe the lens and get what we called a "ballpark quote." Based on the final quote not changing substantially, they issued purchase orders for both lenses.

Monsanto, Fairchild and Hewlett-Packard were interested too, with Fairchild providing prints for quotation. Every call led to new people or companies to see. There was an unbelievable thirst for our capability because up to this point, they had only been investigating making such parts with typical custom molders who didn't know anything about optics. Our status as a provider of lenses to Bowmar and T.I. gave us special credibility. In the next six months I made six more trips to Silicon Valley, bringing home orders each time.

It has always amazed me how much can be learned by simply asking customers and suppliers lots of questions. It also amazes me that so many people are reluctant to ask; either because they don't think about it, or are fearful of revealing that they don't know all the answers. In any event, time and time again the payoff has been phenomenal from asking questions, and never more so than from asking Noah Coughlan what he would do if he had a company like ours.

We heard, in late 1972, that Bowmar Canada's president, Jack Magarian, had become critically ill and was hospitalized for a lengthy period. Sometime during that hospitalization he had been fired, and we heard nothing more.

Shortly before my second or third trip to Silicon Valley, it was announced that National Semiconductor was going to make LED displays and calculators. I called to find out who was heading the program so I could make an appointment. It was Jack Magarian, one of the toughest, smartest, most exasperating and most fun people with whom we ever did business.

Jack had an interesting history. He graduated as an electrical engineer from the Massachusetts Institute of Technology and was employed by one of the first transistor makers, Transitron in Boston. He then moved to Fairchild Camera and Instrument in Palo Alto where he worked with an enormously talented group, including Robert Noyce and Gordon Moore, who later founded Intel, and Charles Sporck, who later headed National Semiconductor. Fairchild sent him to Portland, Maine to run a new semiconductor plant. After a few years, he was hired by Bowmar. He created Bowmar's LED operation, which required finding a Korean electronics fabricator that could wire each segment of the minute LED material to a circuit board. This work was done by hundreds of women being paid a few dollars a day to perform the intricate wiring task while viewing it through microscopes.

Now at powerhouse National Semiconductor, Jack was heading the program with the objective of making it the high-volume, low-cost producer. He was more than just a little unhappy for having been fired while hospitalized with a life-threatening condition. Making life miserable for Bowmar was a secondary objective.

When I called him at NSC he was delighted because although we had not met, he knew and appreciated USPL from the good job we had done for him at Bowmar. At our first meeting in Santa Clara, he requisitioned two bubble lens molds and huge follow-on part orders. It was a tough negotiation, and I took the business at an aggressive price that required maximum manufacturing efficiency. We had a pleasant dinner that evening talking about the business,

Bowmar and our days in Portland, Maine when I worked for S.D. Warren and he worked for Fairchild.

Weeks later, after we tested and qualified the molds, it was obvious I had underpriced the lenses. I went back out to Santa Clara and explained the problem to Jack late one afternoon. He said, "A deal is a deal." I said, "You're right—but do you want a supplier on a low-cost, key part who's not making money?" After an hour or so of intense debate, during which I was terrified as to where this relationship was going, he finally agreed and accepted our revised price. Then he looked at me, smiled, and said, "Roger—why should you and I be little and weak—let's go have a drink and be big and strong!" I couldn't have been happier with the outcome and, at that particular moment, his last suggestion. Jack Magarian not only became a great friend, but he would also become USPL's biggest ever LED lens customer.

Our involvement in an alternative method to magnification for making sizeable alpha numeric displays, called reflector digits, is a bit technical, but is worth relating because it represented an error in judgement. The cost of using lenses to make such displays, over two-tenths of an inch in size, was excessive because too much expensive gallium arsenide was required. To solve the problem, LED makers developed an alternative reflector technique. Each digit segment was molded with a hollow triangular shape in about a 1/4-inch thick piece of white plastic, with the top being the wide readout area and the narrow bottom where an LED spot source was located. When the LED was activated, its light bounced off the internal shiny plastic surfaces and was evenly distributed over the large readout area. A molded red filter cover containing glass diffusion beads was put on top of the digit to further spread the light and be the readout viewing surface. Each digit had seven of the triangular sections. They were made as singles or in various multiples of up to eight. Such digits were considerably more costly than the small magnified ones, but it was the only economically reasonable way to make larger sizes.

We went after orders for the reflectors and covers, even though one could argue it was a stretch that they met our optics specialty requirement for the kind of business we would solicit. I justified or perhaps rationalized the decision on the basis that the interior surfaces of the reflectors had to be highly polished so light could be bounced to the viewing surface.

In the spring, we received an order for eight molds from Fairchild Camera and Instrument. We also received reflector orders from Litronix and Texas Instruments. The molds, with all their highly polished triangular segments, were very complex to design and build. A six-digit reflector display required forty-two segments just for one cavity; typically such a mold would have four cavities, requiring 168 segments. Good custom molders who could not make lenses were able to make reflectors; and, frankly, a number of them were better at it than we were. Our mold-making operation was overflowing with business, so we went to a couple of outside vendors for some of the reflector molds. Having never made such molds, they were not very good at it either.

We muddled through filling these orders, but our costs were excessive and the buyers were not happy with our performance. The worst thing about it was that customers who greatly respected us for our good service and ability to make lenses uniquely, had to wonder if we were really as special as they had first thought. It was my fault. With the benefit of hindsight, we should never have made reflectors because they deviated too much from our specialty and we brought nothing to the customers they couldn't better get from someone else. Fortunately, not too much damage was done and we recovered from the experience. It was a clear reminder of why it was important to be disciplined about the kind of work we accepted.

In March we needed more capital, so I made an infusion of $38,000 in new common stock and final plans were made for the factory expansion.

Early in the year, Don Duff was transferred by T.I. to Italy, where he continued working on calculators for the European market. Soon I started receiving letters from him saying that business was exploding there, and it should be reflected in our orders from Dallas. He was encouraging me to come over to check out the opportunities. I was extremely busy, so a visit would not be possible before the fall.

Ellis and I were getting to know each other quite well as he probed about plastic optic possibilities and helped us with the design chapter of the Handbook. We both knew that USPL had a serious lack of optical expertise. Ron Byrd had left the previous summer, and we had been so busy producing orders and expanding that we didn't have time to fill the position. Also, with all the growth from the LED companies, I was struggling to define the kind of talent we would be needing in the years ahead. One day in the spring, I asked Ellis who the best optical engineer was he had ever met who would be appropriate for a company like ours. He said he knew, but couldn't tell me because the fellow worked for one of his clients and it wouldn't be ethical. I said, "Okay, tell me who he is and I promise I will not go after him based on this information." He said the person was Brian Welham of Bausch & Lomb. Ellis had worked with him when they were there together. I honored my word and did not contact Mr. Welham based on Ellis' information—but there will be more to the story later.

In May, 5,000 copies of *The Handbook of Plastic Optics* were delivered. We ran a few small advertisements for it and had it displayed in the new literature section of a number of trade publications. It was a first-class-looking hardbound book that was excellently printed. It sold briskly at $5 and was given to visitors as well as prospects when making sales calls. There was no doubt the Handbook enhanced our credibility as specialists. Most people thought the content was useful, while others thought if we could publish a hardbound book on the subject we must know what we were doing and therefore, never opened it. One purchaser thought it was a "ripoff," so we returned his $5 and let him keep the book. There were a lot of good comments about the Handbook, but my favorite was from an engineering friend at Polaroid who said, "It's a sure cure for insomnia!"

When fiscal year 1973 ended on June 30, sales were $856,000—up 146%; net after-tax profit was $111,000—up 258%. There was no long-term debt, and shareholder equity of $361,000 was up 136%. Polaroid business had more than doubled to well over $100,000. Semiconductor LED lens volume had gone from nothing to $213,000 with a comparable level of accompanying mold billing. More importantly, our running rate and order bookings indicated that fiscal year 1974 would be outstanding.

# 13 Henry Kloss And His Projection T.V.

At the end of August, 1973, Henry Kloss, the president of Advent Corporation, phoned USPL to make an inquiry. Everyone in the office was at lunch except for Dick Wessling, who took the call. Mr. Kloss asked Dick if we could make a Schmidt corrector plate. Because Schmidt correctors are used in telescopes and our expert telescope maker happened to take the call, the answer was "Yes." If he had called an hour earlier or an hour later, he probably would have been told we didn't know what he was talking about. Who says luck doesn't count in business?

Dick briefed me on Schmidt correctors. They were fairly uniform in thickness with an aspheric surface, which canceled out spherical aberrations in images reflected from spherical mirrors. He felt we could easily make what Mr. Kloss wanted because he had considerable experience in figuring such surfaces.

On October 7, I called on Henry Kloss at his office in Boston. For a moment I thought I was meeting Benjamin Franklin; he bore a remarkable similarity to popular images of Franklin. Henry had created three important companies. He dropped out of the Massachusetts Institute of Technology to found Acoustical Research, the first maker of small, very high-quality acoustic suspension high-fi audio speakers. After Acoustical Research was sold, he founded KLH, also a maker of small speakers that became even more prominent. He decided nothing new was happening in the development of television, so he sold KLH and with the proceeds founded Advent for the purpose of making an innovative television system which projected an image to a large screen. To help finance the projection television effort hereafter often referred to as PTV, Advent also sold a line of improved acoustic suspension speakers.

Henry showed me how he was compression-molding his Schmidt correctors. It was a process that took hours, and the quality was not good. He provided the specifications and I agreed to quote the part quickly. We then went to a demonstration room to see a projector. The picture on its 84-inch diagonal screen was dazzling! At that moment, I concluded that someday projection TV would become a big business. Little did I know then that this was the most important sales call I would ever make, and that it would lead to our company becoming a worldwide leader in an important industry.

# 14 Run-Run-Run: I Need Help

I was doing all the outside selling and being run ragged. Between October 1972 and the end of September 1973 I made nine trips to Dallas, eight to Silicon Valley, six to Boston, two to New York, one to London and one to Canada. In addition, I had to do all the other things required to manage a fast-growing business.

One evening, in a discussion with my friend and next-door neighbor Ward Withrow, I explained that we needed help in finding some first-class talent. He suggested talking to a person his bank had used by the name of Charles L. Arnold. Charlie Arnold had retired as head of human resources at the Kroger Company and was now doing personnel consulting. Ward held him in extremely high regard.

Charlie came out for a visit. We quickly developed a comfortable rapport as I told him about the problem. But, instead of my interviewing him, it was more as if I was the one being interviewed. Within an hour, he knew much about the company and had a thorough understanding of our needs. We decided he would interview three local executive search firms, and we would select one for the task of finding a high-level person to become vice-president of sales and marketing. Charlie then would interview and evaluate the candidates to give me additional input.

We selected the search firm that he thought would do the best job and brought one of its principals, Keith Baldwin, out to meet me. Keith talked about different levels of people he might go after, explaining that there was a wide range of abilities from which to choose. I told him to show us the best broad gauge person he could find. That evening he went home, walked over to his backyard fence and said, "Dave, you told me that if I ever saw a really special little company that might be looking for someone, to let you know." Dave Hinchman was his neighbor.

Charlie Arnold was a wonderful person with incredible human relation skills. He later provided valuable service to us by interviewing and testing people for many high-level positions; giving me insight into the candidates, and the candidates insight into USPL. Everybody thought he was terrific. Tragically, he and his wife disappeared on an annual spring outing to a Kentucky nature preserve. One of them obviously had some kind of a problem and the other tried to help to no avail. They were found days later, but it was too late. It was a very sad time for all of us who knew him.

# 15 David F. Hinchman

In mid-August, Keith Baldwin arranged for me to meet Dave Hinchman at USPL. Within ten minutes I recognized that he was smart, polished and comfortable to be with. Prior to hiring the search firm, I had interviewed a number of sales people who had been referred to us. It was clear this man was in a different league.

Dave grew up in Chicago and went to Princeton, where he graduated with a degree in Economics. We were the same age, had similar military experience, and had both been with large companies. After working a few years for American Cyanamid he moved to Continental Can, where he became sales manager for the Cincinnati plastic bottle facility. His primary customer was Procter & Gamble. I viewed all this as excellent background and was pleased he had selling experience with first-class companies. The plastic bottle business, while far from plastic optics, did involve thermoplastic polymers, molding machines and molds that gave him a good background to learn our processes.

In the initial interview that probably lasted for over two hours, we discussed his background, experience and aspirations. I told him my vision for our future plus everything I knew about our opportunities, challenges, strengths and weaknesses. I felt, and thought he felt, that we hit it off very well.

The next step was to have him spend a day with Charlie Arnold. I explained that he would get as much insight into us as I would into him from this meeting. They met on August 28 for a six-hour plus session. We never received a Charlie Arnold report that was more positive than the one he sent on Dave.

Under the section titled "General Reactions" he said, "Dave is a fine looking, neat, most personable man of thirty-eight. He makes a good first impression and backs it up as he goes along. Hinchman and Howe seem to have good chemistry reactions and have developed good rapport in a short time. I believe they will work well together and that, should any problems of a business or personal nature develop, they can be solved through an open approach and receptivity towards thorough communication."

In a section called "Hinchman's Pluses," he went on to list twenty-six favorable points. In his summary he said, "Howe's initial description was a man who could travel with him and supersede him gradually in a number of years. I am of the opinion that Hinchman's ability to gain acceptance is such that Howe could begin to feel weight being lifted from his shoulders within one year—perhaps even less."

As everyone familiar with USPL knows, Dave Hinchman was a great hire. He has been a major partner in the building of the company, and did a wonderful job leading it to substantial new highs after taking over as CEO in the middle of 1988. In spite of being in many intensive situations together over the years, there was never a time that I know of where we had a serious disagreement. He more than fulfilled my expectations for the top-flight help that was so sought after in 1973.

On September 7, Dave and his wife, Hobey, came to our home to celebrate his joining USPL. It was the same day I met Henry Kloss and first saw a projection TV. That evening I excitedly told Dave all about it. I did not, however, predict the technology would become the principal focus of our business lives in the years we would work together.

Dave Hinchman started his career at USPL as vice-president of sales and marketing on September 10, 1973.

# 16 1973 Continued

With Dave on board, the first task was to get him indoctrinated, both inside and outside the company. We made introductory trips to Polaroid, Texas Instruments and the companies in Silicon Valley. Everyone was pleased with our new player. Prior to this, the only contacts these customers had were Bob Campbell and me. Bob did a great job of relating to them and responding to their needs when I was traveling, even though he was extremely busy running the factory. Dave's presence spread the workload and demonstrated to them some management depth that had been missing.

In October we moved the molding machines from three different locations to the 9,600 square-foot addition. The first molding room that was supposed to take us well into the future had filled up in slightly over a year. As this move was being made, ground was being broken on an 8,000 square-foot addition to greatly expand the tooling operation, which had significantly outgrown its area in a year. One could reasonably question our ability to plan business growth and space, but it was a happy problem. The associates working at USPL now numbered 110.

In response to Don Duff's frequent communications, I went to Europe to see if there was an opportunity for us. Joyce joined me. Calls were made on a number of companies in England and Germany, including the top British calculator maker, Sinclair, located just outside of London, and Texas Instruments near Munich. The trip was very interesting, but the fact was the makers were getting their logic chips and LEDs from American companies. Because of that, we already had all the lens business.

On a side trip to Garmisch, in southern Germany, we arrived one evening to see a huge and frightening headline in the local paper that contained only two words we could understand—"Nixon" and "atomic!" It took us hours to find out that the U.S. had not dropped the bomb and that they were simply speculating about what we might do in an important Middle East conflict that was taking place at the moment.

Don Duff met us in Rome on the weekend to show us the city. He drove down from the T.I. plant in Rieti with an attractive young lady friend who worked there. They came in an older car she owned and had just learned to drive. Rome, on a weekend, is not a place for inexperienced drivers. Nevertheless, the four of us toured for hours in the tiny car with all its loose-fitting parts and her at the wheel. Although no one was hurt, it was a truly scary experience leaving us with upset stomachs. Don quite properly made a separate room reservation for her. The next morning this visibly-relieved young woman made a classic comment to Joyce at breakfast. She said, "Today is the tomorrow I worried about yesterday that never happened."

I had become good friends with Don Duff, our first T.I. customer. We stayed in touch for many years, although he left the electronics business shortly after his Italian posting.

On October 6, Egyptian forces crossed the Suez Canal into the Sinai, marking the beginning of the Arab/Israeli war. It ended with a UN cease-fire on October 24, but the effects were felt long after. The twelve members of the OPEC Oil Cartel temporarily cut off exports, and prices nearly quadrupled in a matter of months. The world market price for a barrel went from $2.91 in 1973 to $10.77 in 1974. At that time, between 1% and 2% of the oil consumed in the United States was used to make plastic. There were severe shortages of some polymers, causing molders to make substitutions to keep their production going. This raised havoc with our supply lines as we were put on allotment by Rohm & Haas and Richardson, the two primary vendors. We were very small users by their standards. The problem was our business was growing rapidly, so we were in big trouble.

I had been through allotment periods in the paper business and knew that frequently it was very subjective as to who got what.

The most important task at that moment was to make sure we had an increasing plastic supply, so I visited the polymer product managers at Rohm & Haas in Philadelphia and Richardson in Madison, Connecticut. The Polaroid SX-70 camera and hand-held calculators were high profile products, so they understood our problems. It was explained that we were not substituting their material for others, that we had a track record of growth with them, and had always paid their invoices within the terms and conditions they set forth. We knew that typical plastic molders were notorious credit problems for them, so this point had meaning.

The net result of these visits was that we were able to get all the raw material we needed and substantially increased our usage at a time others were being cut back. Our policy of paying our bills on time and having a close relationship with key vendors, paid off then and would again in the future. It is common for companies to extend supplier payments as far as possible beyond normal agreed-upon terms. Although some think it is naive, I always insisted on paying our invoices when we said we would pay. Because of this, there was no doubt we received preferential treatment at critical times, far outweighing the benefits of pushing vendors so their money could be used to run our business. It also seemed to me to be the ethical thing to do.

In November, Advent introduced the Model 1000-A projection television with three Schmidt corrector lens plates made by USPL, representing about $12 of content. The unit was in two pieces, consisting of a projector and a 7-foot diagonal screen. The selling price was $2,500. We thought the quality was quite good, but there was nothing to compare it with. Most significantly, it was the first consumer projection television, and it made TV makers all over the world sit up and take notice.

As 1973 came to a close, we were very pleased with our progress. The only remaining strategic negative was that we had not brought in a high-level optics person. We were worried about a recession in 1974 causing new equipment purchases to be slowed.

Two year-end memos I wrote tell a lot about our thinking at the time. One was to all associates, and the other was to the management group regarding our being a specialty company.

TO:   The People of U.S. Precision Lens

DATE:   December 3, 1973

SUBJECT:   U.S. Precision Lens In 1974

We have all been hearing and reading about the energy crises, recession, unemployment, rationing and material shortages. These are things that could happen, so I know you are concerned about how it all could affect our company. I cannot (and nobody else can either) tell you what is going to happen, but I can tell you what we expect.

Regardless of what happens to the rest of the economy, for our company to do well four things have to happen:

1. We must have orders.
2. We must have material.
3. We must provide good quality.
4. We must control costs.

Let's examine what we are doing on each of these points.

OUR ORDERS—We have talked to all of our major customers and have analyzed our orders carefully. We have as high a backlog of unfilled orders as we have ever had. Our customers tell us everything looks up for 1974. The calculator business is forecasted by some people to double. We have no idea whether that will happen, but many new low-priced models will use our lenses and our customers think the demand will be big. The Polaroid SX-70 is selling so well that they claim they cannot be made fast enough. Our sales to them should be even bigger than in 1973. Our general lens business is continuing to grow, too.

RAW MATERIALS—Between 1% and 2% of the oil in this country is used to make plastic. For that reason, there is a real plastic shortage. Our main plastic—acrylic—is in short supply, but not as short as some of the others, such as styrene. We are told by our suppliers that they will take care of our needs. We have placed orders months ahead of time and, as of now, have no reason to think we will not get what we need.

QUALITY—With our customers, we have the reputation of being a very high-quality producer. A few months ago, I told you we had to do a better job because we were getting too many complaints. We have obviously worked hard on this because the complaints are down substantially. If we keep up the good job, we should have no fear of losing orders over quality.

COSTS CONTROL—We are doing a good job of eliminating the inefficiencies that came with overcrowding, problem molds, and new conditions, as a result of the move. Everyone is working very hard on this problem and the results are really starting to show. In the last few days, we have had the best production ever in the Plastics Department. It is greatly appreciated. Because of the doubts about the economy, we have either canceled or delayed new equipment purchases. This will allow us to stop borrowing money from the banks over the next few months and pay off some of our debts.

When you add all this up, things simply do not look that bad for U.S. Precision Lens. Of course, if our customers cannot sell their products they will cut our orders, but we see no signs of it now. Our raw material suppliers say they can take care of us; but if they should have trouble with their suppliers, we will feel it. Again, we do not expect it to happen. We are doing a good job with quality, and production efficiency is improving nicely. When you have that going on with a slowdown in equipment spending, we should be getting in a sound position for tougher times if they should come.

We have one thing going that will serve us well when the chips are down. Our tooling and polishing people have the reputation of being tops in their field and very responsive to customer needs. We have no competitor that can offer that combination. Frequently, it is that difference that gets us new orders.

As I said before, no one really knows what is going to happen, but I'll tell you what I think we will see. I believe there will be fairly high unemployment, but not at U.S. Precision. There will probably be a mild recession, but not a bad one unless the Arabs do not loosen up on oil. Finally, I think we are going to get the material we need to run full. If anyone has any questions on this, by all means have your supervisor ask me and I will do my best to give you an answer.

Best regards.

*　　　　*　　　　*　　　　*

## December 26, 1973
## THE STRATEGY OF U.S. PRECISION LENS
## AS A SPECIALTY BUSINESS

For the last three years, we have been talking about USPL as a specialty business. We have planned and worked to position the company as a specialist, and it has paid off beautifully. Because I feel this is the single most important USPL business concept to understand, I will review it. Also, I will try to put on paper where I think we stand with respect to it, and what we have to do to insure it staying with us in the future.

## THE SPECIALTY GROWTH COMPANY

There are specialty companies with small markets that are not growing. This discussion does not refer to them. Specialty companies have the ability to get ahead of competition in skill, market position and profitability. Because they have been able to do these things, it naturally follows that they are generally more responsive than their counterparts. Their customers regard them as out of the ordinary because they know how to provide things that cannot be easily obtained from others. These companies are more profitable because they can charge a premium for their products. Investors pay a premium for specialty stocks because their profitability and lack of competition allow them to grow at a faster rate and, therefore, compound the advantage. We have seen this phenomenon clearly with our own company. We use similar equipment to custom molders, we use a simpler array of materials, and we use molds that, in all but one respect, are comparatively uncomplicated. In spite of this, we have figured out how to sell our machine time at two to three times the price of the nonspecialist.

## HOW WE POSITIONED OURSELVES AS SPECIALISTS

We convinced customers that we were an optic company, not just a sophisticated molder. We worked hard to get the equipment and skill necessary to be different. We grasped every usable (if not usable—saleable) straw to convince our customers we were unique. Finally, we broadcast this through publicity, advertising, and our book. In all of this, we were responsive to the customers and their production needs—a point that cannot be overestimated in importance.

## USPL AS A SPECIALTY COMPANY TODAY

We are unquestionably known as the supplier to the open market that is best able to furnish customers their optic needs. American Optical is sleepy, and Polaroid, Kodak and Bell & Howell are not interested in outside customers. Diverse Technology has the concept, but not the resources or market position. As for us, we are better than ever, but we have some cracks in our armor that need to be filled. We still do not have good ways of measuring and talking about quality specifics. We should be able to help our customers technically better than we presently can. We should be able to talk with them about more complex optical systems knowledgeably. We should make a real effort to establish ourselves as aspheric experts. The time is near to renew our promotional efforts so our name is more frequently seen.

## WE CAN EXPECT COMPETITION

A business like ours attracts a lot of attention, which naturally invites competitive examination. The higher cost of glass optics, plus our promotion of plastic, is getting people excited about what we do. Diverse Technology is going to chip away at us and could merge into someone with the commitment and resources to back the concept. Then, they may become a formidable competitor. Bell & Howell, for whatever their reasons might be, could become a tough competitor because we would be vulnerable technologically. American Optical could wake up with just a slight management change. Other glass optic companies may conclude they have to get into plastic to protect their markets.

## ONE SIMPLE CONCLUSION

If this assessment is true, our job is obvious. We must work on making the specialty more special, more mysterious and, therefore, more difficult to get into. Considering where we are now and the present lack of effective competition, I think doing this will be relatively easy. It will cost some money and will require concerted effort, but I do not see anything about it that should baffle us. It is a top priority along with keeping the business financially healthy.

Remembering where we were three years and two and a half months earlier when this adventure began, there had been a phenomenal amount of change—but nobody was looking back because we had wonderful opportunities before us.

# 17 1974

World calculator production was growing phenomenally. It had gone from 3.5 million in 1972 to 21 million in 1973, and would climb to 34 million in 1974. There were probably over a hundred calculator makers in the world, and they were coming into and going out of business with regularity.

Only a few were integrated with the ability to make the logic chips, the LED displays, or both. Those companies were Texas Instruments, Rockwell International, National Semiconductor, Litronix, Bowmar and Hewlett-Packard. All others had to buy the two key components, placing them at a significant disadvantage. T.I. and NSC would sell both components to anyone. Rockwell and others sold logic chips, and Litronix was a major LED display supplier to foreign markets.

Commodore was the largest of the assemblers, and even though it had to buy the components, it was an aggressive leader in driving down the prices because of a willingness to accept razor-thin margins. Hewlett-Packard made only scientific and advanced business models, which put it in a special class by itself. In 1974 H.P. sold 300,000+ units, while T.I. was in the area of 4,500,000 to 5,000,000.

Prices were dropping like a rock. In 1972, the T.I. basic four-function instrument sold for $150. By August of 1974, it was $45. Rockwell had models as low as $30. NSC and Commodore had stripped down devices for $20.

Our LED lenses always sold at prices from 4.5 cents to 9 cents with, of course, the customers paying for the expensive molds. The simple early versions cost about $10,000, increasing to the $20,000 range for more sophisticated designs. Usually we had 5 cents to 6 cents in a calculator, plus 3 more cents if a viewing filter window was included.

The biggest opportunity for cost reduction was to miniaturize the expensive gallium arsenide. As this happened, higher-power bubble lenses were required to keep the digits at a proper viewing size. The problem was as the lens power increased, viewing angle was undesirably decreased. To overcome this, there was a progression of new advanced designs. Either the LED makers designed them or hired Ellis, or we had Ellis do them.

The first cylinder lenses, lenses that only magnified in one direction, quickly gave way to spherical bubble versions with flat back surfaces. When the power of these was pushed to the limit and caused unacceptable viewing angle, spherical meniscus designs with strong bubble power but negative bottom power provided improvement. During the next three years there were further dramatic improvements with complex aspheric bubbles and two-piece, bubble-cylinder combination designs. Throughout all of these advancements our prices remained steady, while our customers were benefiting from huge manufacturing cost reductions in their processes as a result of our technology. What we got out of it was very profitable enormous increases in volume.

A new LED opportunity was rapidly developing for digital watches. In 1969, SEIKO developed the first electronic watch. It utilized a quartz crystal oscillator and a semiconductor circuit, which made for a simple design that was accurate to within five seconds per month. In 1972, Pulsar introduced a $2,100 model that had an LED display.

With all the gears and electro-mechanical parts replaced with mass-produced electronic devices, most of the labor and material had been designed out. The electronic companies saw watches as a big opportunity. Liquid crystal displays could also be used but were not very good at that time, so LEDs became the dominant readout technique.

In 1974, with worldwide watch production at 205 million units, only 700,000 were digital. It was predicted that digital models would grow to 50 million in 1980, causing many companies to get into the business. Litronix, Texas Instruments and National Semiconductor had introduced or were planning to introduce models, and were selling components to assemblers. NSC later introduced LED versions that sold for as little as $22.

LED watchmakers not only had the same expensive gallium arsenide problem that was present with calculators, but they were also concerned about power consumption from the small batteries. Again, they looked to display miniaturization with magnification from lenses as the answer. Some of the watches were made with lenses that were directly cast over the LED. Many were made without magnification, in spite of the power consumption and cost. We produced watch magnifiers for a number of companies and it was good business, but never approached calculator volume. LED watch growth failed to fulfill the hopes of the optimistic forecasters. In time, liquid crystal displays would improve and eventually take over the market.

The disposable camera, so common now, is not a new idea. In 1972, we had a customer in Philadelphia that was producing one using a Kodak Instamatic film pack as its back. It had only modest success. In 1974, a small outfit on the West Coast had the same idea. While calling on the two Los Angeles fellows who ran it, I discovered one had been an engineer for Bausch & Lomb years earlier. At lunch I asked him who B&L's best optical engineers were. Among those he mentioned was Brian Welham. This was the person Ellis told me about and I agreed not to recruit as a result of his providing the information.

Immediately after returning to Cincinnati, I called Ellis to tell him what happened. I said that with the West Coast B & L alums' recommendation, I felt released from my commitment. He agreed, and I called Brian right away. Soon thereafter he came to Cincinnati and we had an introductory dinner. This was followed by a visit with his wife, Jennifer. Dave, along with his wife Hobey, Joyce and I introduced them to Cincinnati and USPL in depth.

Dave and I were sure Brian would be superb for USPL's needs. We made an offer which we thought he accepted. Shortly thereafter, he called to say he would not be coming because he couldn't leave an important project that had to be completed. We were extremely disappointed. I asked him whom he could recommend for the job. He gave us a few names, but the follow-up interviews didn't reveal anyone for whom we had enthusiasm. We continued the search for the critical optical engineering position. However, it was not going well.

Fiscal year 1974 ended June 30, with sales of $2,364,000, up 176%. Net after-tax profits were $257,000, up 132%. Long-term debt was zero, and stockholders' equity was up 71%. The LED lens business jumped 293% to $836,000; Polaroid increased 256% to $332,000; and all other lens customers, which were lumped together in a category called Components and Systems, more than doubled to $485,000. This could not have been done without large mold sales. The economic conditions that caused worry at the end of 1973 did not materialize in a way that hurt us. With Polaroid continuing to do well, and the LED business exploding, the future looked terrific.

In an effort to upgrade our process capability, we were searching for an engineer who understood molding. In September, Bob Campbell interviewed Donald L. Keyes. Bob was unusually enthusiastic. Trained as an electrical engineer, he was one of the first employees of Hunkar Laboratories, a Cincinnati company that made process control equipment primarily for the plastics industry. Don was an expert on plastic materials, molding theory, and how to program injection molding machines. After meeting with him, I was also impressed. To our delight, he accepted our offer and joined us on October 10.

Not counting our brief earlier association with the optical engineer, Don was our first highly-trained engineer. Over the years, he was the company's material and processing expert. He also consistently made important contributions in a variety of technical areas, including suggesting to me an idea that was a critical element in a breakthrough lensmaking process. It was USPL's good fortune to have Don Keyes as one of its builders.

As calculator prices plummeted, an accounts receivable problem developed with Bowmar in September. They had just built a state-of-the-art logic chip plant in Phoenix for $7 million, which led to a severe cash shortage. The pain it caused was being passed on to their suppliers. Although owed roughly $60,000, we were not nearly as stretched as their other vendors because with a 6-cent key component without which they would shut down, we were able to extract preferential payments.

To our complete shock, we learned that the term "preferential payments" also had a legal connotation. At the time, if a company was determined to be technically bankrupt even though they had not yet filed for bankruptcy, payments made to vendors ahead of those made to other vendors were recoverable. In the United States, when a company declared bankruptcy the court could recover the last three months of payments—and in Canada, the last four months if it could be shown there was preference over other creditors.

We were deeply concerned because with Bowmar being a heavy purchaser, we had $200,000 to $300,000 of their money that could be construed as a preferential payment. At $300,000, half of our net worth was at risk. It was fairly obvious that without a miracle they were headed for bankruptcy, and it would be soon. Dave and I, along with our lawyer, Don Lerner, went to Bowmar's New York headquarters to get our $60,000 and learn as much about the situation as possible.

That summer, the Franklin National Bank of New York failed. It was one of the biggest United States bank failures ever and had been important news for months. As we pulled up to the Bowmar address, I said to Dave and Don, "This is a very bad sign." Bowmar was in the Franklin National Bank building.

We were ushered into their extravagantly plush offices. It looked as if someone had challenged the decorator to see how much money could be spent. The thirty-something treasurer greeted us. As he led us to a beautiful conference room, I realized he had been profiled the week before in *Business Week*. He and other Harvard Business School graduates were interviewed extensively about career paths and what they were doing with theirs. With half our net worth at stake, we were worried about ours too!

In a long-winded dissertation, he proceeded to explain how the calculator business worked, what was going on with it and why Bowmar would be successful. It was simplistic and had nothing to do with why we were there. Finally, I said that his presentation was very interesting, but we simply wanted our money or we would have to stop shipping lenses. He slipped back into his tutorial, and we patiently waited for it to end. We asked for the money again. Showing considerable exasperation, he said, "Would you take a great company like Bowmar down over a 6 cent part?" After a brief silence I said, "If a little company like ours could take a great company like Bowmar down over a 6 cent part, we would rather know it now than later!" As we left, without the money, we again emphasized that Bowmar was not going to get any more lenses unless we were paid.

A few minutes after leaving this fellow, the day got worse. We called our factory to make sure no Bowmar production was running and learned that Texas Instruments called to instruct us to ship four calculator filter window molds to their Lubbock, Texas plant because, with a temporary business slowdown, they wanted to fill in-house excess molding capacity. These molds made parts that were consistent with our specialty, but just barely. They did represent substantial volume.

As it turned out, with a lot of luck we got out of the Bowmar problem unscathed. Our customer was Bowmar Canada, a wholly-owned subsidiary. Its lawyer was on the board of directors and because of some rule of Canadian law, he was personally liable if it did not file for bankruptcy but was, in fact, technically bankrupt. With that important bit of information, we readily received his legal opinion that Bowmar Canada was technically not bankrupt. In addition, the receivable was paid because they had to have lenses to keep operating.

We instituted a cash-in-advance requirement for their future orders because it didn't count as a preferential payment. As predicted, the parent, Bowmar, went into bankruptcy and by the time the Canadian subsidiary had its bankruptcy problems, our four-month risk period had passed.

Today, the numbers involved do not seem that large, but then, a bad outcome of the situation would have been devastating. It was sad to see our first LED customer fade away. With the calculator business so brutally competitive, Bowmar was its first major casualty.

Litronix had substantial LED volume throughout Asia, plus manufacturing facilities in Singapore and Malaysia. All our lenses were shipped to those two plants for assembly. We worked very closely with Dan Davis, a senior engineer and product manager. Dan encouraged me to join him on a trip to that part of the world because he thought we might find some opportunity. He further suggested we take our wives, and I readily agreed.

It was my first trip to the Orient, and proved to be extremely educational. Our first stop was in Japan, where we met with Rikei Corporation, the trading company that represented Litronix. They set up sales visits to various users of lenses in Tokyo, Kyoto and Osaka. The Japanese were quite interested in our technology, but no promising leads came out of it. We both benefited by gaining a much better understanding of the Japanese calculator industry and market.

In Korea, while Dan was visiting the Litronix trading partner, I spent time with a Texas Instruments engineering friend who had been sent there to oversee a huge contract assembler of LEDs called

ANAM Industrial Company, Ltd. Jack Magarian first got ANAM in the business of doing the intricate wiring of the circuit boards to the LED material when he was at Bowmar. Now, with Bowmar leaving the picture, the factory's expanded capacity had been taken over entirely by T.I. and National Semiconductor. When I went through the production areas, white USPL boxes were everywhere. ANAM was the largest assembler in the world, and many millions of our lenses were shipped there to be attached to calculator display boards.

Having heard great stories about Hong Kong, when we arrived I found it even more fascinating. Litronix did an enormous business through its trading partner there, Astec Industrial. While Dan visited his customers, I went to see makers of calculators and watches. Again, we were getting all of the lens business through the U.S. LED makers. Nevertheless, much was learned about doing business in that area from Astec, and particularly from a meeting with a Bank America manager.

A couple of days were spent in Bangkok, and we were supposed to go to Taiwan, but lost our air reservations in a confirmation mixup. The last stop was in Singapore, where I was to visit Litronix and Hewlett-Packard factories.

One of the Litronix engineers I knew had been sent to the Singapore operation a year earlier. We visited in California a month or so prior to the trip, at which time he told Dan and me that he was marrying a Singapore girl of Chinese descent. He invited us to the wedding. We told him our schedules did not fit his timing, so we had to decline.

When Dan, his wife Madeline, Joyce and I arrived quite late in the day, we asked the hotel to direct us to the best Chinese restaurant in the area. The minute we entered, we inadvertently found ourselves at this fellow's wedding reception. The beautiful Chinese bride's father had purchased a large amount of land in downtown Singapore right after World War II and now had phenomenal wealth. The reception, first class in every way, reflected his position. As we stood there watching all the interesting people from two different worlds, Joyce and I were tapped on the shoulder and asked to participate in an old Chinese wedding custom where the bride, groom, and the people that have come the farthest joined each other on the stage for a brief ceremony. There was no turning back now so we naturally agreed, not knowing what was going to happen. The procedure was that the four of us would lock arms and each eat a slice of what appeared to be a nearly-raw suckling pig, held by a waiter on a spectacular silver tray. Joyce looked at me with terror in her eyes. With Litronix being one of our largest customers, and the front two rows next to the stage lined with distinguished older Chinese family and guests watching intently, I looked right back at her and, without moving my lips, said, "Eat it!" She did, but it will forever be etched in our memories. We discovered the next day that one can go around the world either way from Singapore, and to New York it's almost exactly the same mileage. We, indeed, had come the farthest.

The trip, with our friends the Davises, was a terrific education and introduction to the Far East. Although no new business was generated, it provided extremely useful background for experiences we would have a few years into the future.

In November I sent the following memo to all our associates, describing the trip and my outlook for the near term. It is fairly representative of the way we tried to educate and inform everyone about the forces that were driving the business. We also had frequent company meetings to keep them current on issues that came along.

November 13, 1974

TO: The Employees of U.S. Precision Lens

As many of you that are used to regularly seeing me know, I haven't been around much lately. Except for some recent family medical problems, which seem to be resolving themselves, the reasons for my absences also affect you, so I thought you might be interested in what I have been doing.

From the middle of September to October 5, I was in California once, Arizona once, Texas twice and New York once. The reasons are simple. With the economy as soft as it is, we are having to work harder to get orders. One customer is planning to remove a few molds, so a lot of time has gone in trying to change his mind. He wants to run the molds so he can keep his people busy. We have other customers that are financially sick and have difficulty paying their bills. This means that much time has gone into trying to get paid for what we have done. The short of it all is something you understand well—inflation and the uncertain economy is hurting everyone—you, your company, our customers and their customers. I spend four to five times more hours on these problems now than I did a year ago. Also, much more of my time goes into the question of what new areas of plastic optics can we get into.

From October 5 to October 21, I went halfway around the world to Japan, Korea, Hong Kong, Thailand and Singapore. The reasons for this trip were as follows.

1. Many of the electronic products that are sold to the world come from these areas. I went to explore specific opportunities for us to make lenses here and sell them to companies headquartered there.

2. Much of USPL's production is sent to plants there for final assembly. Bowmar, Texas Instruments, National Semiconductor, Hewlett-Packard, Litronix, Fairchild and Monsanto all have their LED operations over there. I wanted to see if any competition to these companies or us might be developing. Also, I wanted to meet a few of the key people in plants where our parts are used extensively to find out how they are doing.

3. My last reason was that I wanted to learn about how products are sold there so we will be ready when opportunities present themselves.

I won't bore you with all the details of what I learned, but some may interest you.

## Japan

When most of us think of Japan we think of low cost goods that are not bad in quality these days. A few years ago what they made was low cost and poor quality. They have been particularly good at making electronic products. Historically, they have run their businesses on a "cradle to the grave" basis—by this, I mean companies paid very low wages—people were never laid off and they hardly ever changed jobs. That worked fine for the government and companies while everything was going up, but some things are happening to change it all. Workers wanted more of a fair share, so wages have gone up greatly; the oil crisis hurt Japan much more than us because all of theirs is imported; their inflation has been twice as high as ours. This has led to Japan losing much of its competitive advantage. They cannot make calculators now as inexpensively as U.S. companies, and that's *a very significant change*. I think the Japanese economy is in very bad shape and will get worse. This helps us on one hand, because they are competitors to U.S. industry, but it hurts on the other because they are also big customers. For example, Litronix and other LED makers sell a lot of products using our lenses in Japan.

Incidentally, the Japanese are very interested in our music, dress and ways of doing things. Right now they have gone golf crazy. It is considered good for status to have U.S. made golf clubs. There aren't that many places to play, and they have to make a reservation a month ahead of time. A round of golf costs $50, compared to $5 in Cincinnati. The salesman I made calls with also plays the guitar and banjo. His real specialty is Bluegrass music, and he knows more about Kentucky and that kind of music than probably anyone in our plant.

## Hong Kong

The quality of goods that come from Hong Kong is not as good as Japan, but labor is very cheap. The average daily wage is $3.50.

Working conditions are terrible—so bad I wouldn't know what to give you as a comparison. We drove thirty miles north and looked over the border at Red China. It was guarded by British soldiers.

## Korea

I visited the plant where most of our T.I. and Bowmar lenses are assembled. The average wage in Korea is about $2.00 per day. The working conditions were extremely crowded. People putting LED's together were sitting so close they were nearly touching. I was told these conditions were better than most places there. Imagine putting 120 people in our lunchroom—that's what it was like. I was in Seoul, which is the capital and a city of about 4,000,000. The pollution in parts of it was so bad you could nearly cut it with a knife.

## Bangkok, Thailand

This is supposed to be one of the most beautiful cities in Asia, but I guess I was in the wrong parts. Many people live along canals which were flooded when we were there. The canals are used for *everything*. We took a boat ride to see a particular temple. As we were getting back into the boat we saw a dead man floating in the water. He had obviously been in a fight—nobody seemed to be concerned about it.

## Singapore

This is a beautiful city. National Semiconductor, Litronix and Hewlett-Packard all have plants there. Although the working conditions were poor compared to American standards, they were not that bad. The average wage is $2.00 to $2.50 a day but going up rapidly. At Litronix I was told we give them fewer quality problems than anyone else they do business with. After taking a little beating on that issue in Korea, their comments were nice to hear.

In every place I visited, inflation is worse than it is in the U.S. This means their costs are going up faster than ours, and it will be easier for us to compete in world markets. With the exception of Japan, however, the difference in labor rates are so drastic compared to ours that it will be a long time before the gap narrows significantly. We are able

to compete to the extent we do because Americans are more productive (by that I mean we get more done in an hour), we are more automated and we simply know how to do things that other people can't
do. Although it was an interesting trip and I learned much that should
pay off for us in the future, it sure was nice to get back to America. I
wish everyone could have been with me because no matter how tough
our problems are, one needs to have a comparison to realize just what
a great place our country is.

*    *    *    *

At the beginning of this letter, I told you a little about what was taking
my time before leaving the country. We all know there are business
clouds that make the future somewhat uncertain. We are constantly
trying to figure out what they mean to U.S. Precision Lens. While there
are segments of our markets that look weak, there are new things coming along that may offset them. Considering the economic conditions,
at this point I do not think things are going to be bad for us, although
1975 will probably be a year of no growth. Unless there is some drastic
change in the meantime, near the end of November I will write a letter
on this subject so you know just what we think business will be like in
the early months of 1975.

On November 14, Dave Hinchman was promoted to executive vice-president—just fourteen months after joining us. I was delighted with the job he was doing. Our skills were complementary,
and the entire organization found him compatible. Internally, at
higher levels, there wasn't much sensitivity about titles. The purpose
of the move was to indicate to our customers, vendors and work force
the increasingly important role Dave was playing in the operation of
the company.

# 18 Nobody Wants a Sole Supplier

Because of USPL's nearly unique position as an independent maker of plastic optics, we were a sole supplier to all of our customers. Engineers worried about information security, since we were providing similar lenses to all their competitors. Purchasing people in particular resented not having alternatives. It was a subject that would come up again and again. They had many recurring fears: What if USPL had a fire or a tornado hit? What if it goes out of business? What if there is a strike? Will it have enough production capacity? How do we know there isn't price gouging?

Dave and I decided we had to deal with the issue in a forthright, positive way. If it was brought up, or we even suspected it was bothering a customer, we took the initiative by making these points.

1.  Nobody wants a sole supplier. We understand that because we don't like to be placed in that situation with our suppliers. However, although not our fault, we had this problem with all our customers and, because of it, we deeply believed we had a special obligation to prevent anything from interrupting their getting our products as they desired them.

2.  Regarding security, we knew that if there was ever an unethical act on our part of passing along a customer's confidential plans or information, we would be ostracized throughout the industry. We told them it was vital to our best interests to be above reproach on this question, and there had never been an instance where there was a breach. We consistently promoted the fact that we could be trusted.

3.  Our factory was sprinkled and had firewalls in appropriate places, so we were not concerned about a fire. It seemed to us a tornado was a remote risk.

4.  With respect to business soundness, while not going to the point of furnishing statements, we explained that financially

we were very conservative, paid our bills on time, and had a strong balance sheet. The offer was made to put them in touch with our bankers for confirmation if that would provide comfort.

5.  Because of our sole source position, and uncertainties due to fast growth, our production planning policy was to have 20% or more excess capacity.

6.  Although our business was considerably more complex than the normal custom molder's, we argued lens prices should be compared to the prices of similar-sized plastic parts, and they would find the premium not to be excessive.

By exhibiting understanding of the issue and forthrightly talking about it, for most customers the problem went away or at least was given low-priority status. In 1975, however, it was pressed very hard by Litronix and, although less so, by Hewlett-Packard. They wanted us to have two manufacturing locations so if there was a problem at one they would be covered in the other. Also, they wanted us to have a facility closer to them. After considerable resistance, we finally agreed to study the idea of a California plant.

I collected a lot of information about possible locations. Dave and I decided that if we were going to do it we didn't want to be too close to the customers, for fear they would get overly familiar with the details of our operation. We finally settled on the selection of the Carmel-Monterey area, or north of San Francisco in Santa Rosa.

We had a long, intense board meeting debate over the prospect of a California plant. Don Lerner and Jack Roy had serious doubts as to its wisdom. Their major concerns were the amount of management time it would take and cost effectiveness. I argued the case in extensive detail and they finally relented. The next step was to find a location.

I visited the Carmel-Monterey area and Santa Rosa looking at sites, finding out about building costs and checking labor availability.

The director of the Carmel-Monterey Chamber of Commerce asked how many employees we would need. I said about thirty initially, and he exclaimed, "Oh, that's wonderful—but we wouldn't want you to ever have more than fifty." That certainly was different from what one hears from other area-development executives, but Carmel-Monterey was unique in not wanting large industrial employers.

We continued to study the matter. It was decided Bob Campbell would move there to run the plant because he understood all the details of our operation, plus he was quite effective in dealing with customers. Dave made a trip to look at locations so we would have an additional point-of-view. Throughout the investigation, I often thought about the concerns Don and Jack expressed at the board meeting. That session caused us to look at the feasibility of it in much greater detail and, again, question if it was really a smart thing to do.

As the study was progressing, Litronix's business softened significantly. Now, rather than pressing for a second plant, they were more concerned about cutting back on orders. A California location quickly became a non-issue, and the new priority was to persuade us to extend receivable payment schedules. We told Hewlett-Packard we studied the idea and it simply didn't make economic sense. Our California plant idea rapidly faded away.

At the time, and often since then, I have thought this was a superb example of why an outside board of directors is important. By challenging us as they did, our more thorough investigative process probably prevented a mistake that could have been immense.

The sole-source-of-supply problem would always be with us. The response developed in those early years would be used again and again.

# 19 1975

In January, at Dan Davis's urging, I attended the Consumer Electronics Show. It was my first. In those days there were winter and summer shows, both held in Chicago. A multitude of companies exhibited their latest TV sets, radios, audio components, calculators, watches and just about every other electronic gadget imaginable. Of course, our interest was calculators and watches. All the makers had their top salespeople showing the latest array of products. It was an excellent opportunity to compare everyone's offerings and, best of all, talk to sales executives about market trends. Normally there was no contact with these individuals, and by spending some time with them we gained a more complete view of what was going on in the business. It also gave us a chance to meet with customers and get their reactions to what was seen and heard.

In the years ahead, the winter show was moved to Las Vegas. Our attendance at the two events became a significant part of our sales effort. By lining up meetings, lunches and dinners, we could see many important clients as well as potential customers in a short time.

In April, John M. (Buzz) Bullock, a senior vice-president and loan officer for the First National Bank (now Firstar Corporation), joined the board of directors. We became acquainted in mid-1971 when he called on me. At the time I purchased the company, the banking relationship was with the First National but was moved to the Central Trust Company, which is now part of PNC Bank Corporation, because loan officers there had been the most helpful during the search for a business. Changing banks, lawyers, accountants and insurance brokers was something I didn't believe should be done unless there was a very good reason. After a couple of more sales calls, Buzz impressively said he was going to keep coming until the First National won the business.

All borrowing at the Central Trust Company had been per-sonally guaranteed by me. Near the middle of 1973, I told the loan officer that our balance sheet was now sufficiently strong that, in my opinion, there should be no guarantee required. He shocked me by saying, "If you don't believe in your business, why should we?" I em-phasized that we did not want to change banks; however, confidence had been lost as to whether the arrangement was competitive. To check this, I outlined what we wanted and asked Central, as well as other bankers, to respond and, in addition, invite them all to use their imagination as to what might be even better for us. The door had been opened. Central came back with exactly what we asked for, dropping the request for guarantees. First National did likewise, plus adding meaningful enhancements. As a result the account was moved, beginning a long and mutually-beneficial relationship for both our firms.

Buzz Bullock was an excellent director and helpful in many ways for the eleven years he served us. We never had the slightest doubt about getting first-class service or highly competitive rates from the First National Bank. In 1982 I was asked to become a member of its board of directors, and today I remain on the successor company board, the Firstar Corporation.

Advent's fiscal year ended in March, with a loss of $3 million. The Model 1000A, with our Schmidt correctors, was selling fairly well but was too costly to build and therefore not profitable. To solve the problem, Henry Kloss was starting to develop a new model which utilized red, blue and green 5-inch monochrome television picture tubes—hereafter also called cathode ray tubes or CRTs—that would magnify the TV image through three five-element lenses to a large screen. Heavy expenditures were being made to support the project.

The lenses had to be quite fast, meaning highly light-transmissive; otherwise too much image brightness would be lost, resulting in a dim picture. Henry hired David Gray, a prominent Boston lens designer, to do the design work. When fast glass designs were found to be prohibitively expensive, Henry had him do a plastic aspheric version. He asked us if we could mold the large elements that were over an inch thick and nearly five inches in diameter. It was an extremely difficult molding challenge, so I suggested that a prototype, in single cavity molds, be done on a best-effort basis. That meant we would try our best, but if it didn't work he would still be obligated to pay us for the attempt. He agreed to the condition and our $30,000 price, which was calculated to cover only our direct costs. Overhead, profit and a cost overrun that eventually materialized were our contributions to the project.

In the fall, molding trials were run with two significant problems. The large aspheres did not have enough surface accuracy due to excessive shrinkage of the plastic during cooling, and the molding cycles were from over twenty minutes to as high as forty-five minutes. These were unheard-of long cycles in plastic molding. Even if we had been able to get the surfaces accurate enough, the number of molding machines needed to produce large quantities would have been astronomical. Henry's optimism for the new approach was infectious, and I concluded that if we could find another way to make what he wanted, there was potential for a very large business. We continued to work on the problem by searching for alternative manufacturing techniques.

In September, while waiting in San Francisco's prestigious Clift Hotel for a Litronix group to arrive for dinner, I wandered into the gift shop. A 4-inch diameter clear plastic paperweight caught my eye. It looked like a lens and, in addition, had embedded half dollar, quarter, dime, nickel and penny coins. The price was $13.95. I concluded if the Clift's high-priced gift shop could sell something that so closely resembled a lens for that price, we ought to be able to find a way to solve this problem. I purchased the paperweight and still have it today. Finding that object provided inspiration to try harder.

Brian Christopher and Neil Stewart, the principals in the Hong Kong Trading Company, Astec, believed we had calculator lens opportunities there and in Taiwan. We had an agreement that made them our sales agent for these areas. They wanted me to visit to help get the sales effort going, so in August I made a seven-day trip. There were a number of companies talking about making LEDs, and one in Taiwan actually was doing it. We had shipped an order to that company, so I called on them in Kaohsiung, the second largest city in Taiwan at the southern tip of the island. At the end of the visit my conclusion was that there were good but not great prospects in the area. Unless people in that part of the world could displace Texas Instruments, National Semiconductor and Litronix, the opportunity would be limited.

Before going on the trip, Jack Magarian said he wanted to see me when returning. We arranged to have a morning meeting on a Sunday at the San Francisco airport where I was to arrive the night before. He was very curious as to what kind of competitive threat the makers there would be for him, and how we were going to deal with them. After telling him it was an opportunity we would pursue, he expressed considerable disappointment. He went on to say the USPL LED lens technology gave U.S. LED makers an important competitive advantage that he hated to see lost. Furthermore, he calmly said that if this was going to be the USPL plan, NSC would have to aggressively develop alternative lens sources.

NSC was our largest customer and purchased more lenses than we could have hoped for, from all the Far East makers combined. After a moment of intense silence and with his last statement clearly in mind, it was my turn to speak. I calmly told Jack USPL had rethought its plan and just changed its view. In light of new information, it would not be pursuing calculator lens business in Asia. He smiled, said our company made a good decision, and we parted. If the meeting had taken place earlier, there would have been no need for the trip. It did, however, reiterate to us just how important our lenses were to the LED industry.

Jack Magarian was a colorful manager who personally involved himself with every aspect of the NSC LED operational and sales effort. On one afternoon visit I sat in his office for over three hours as he took one call after another and talked to people who constantly poked their heads in the door. When closing time came, we had not discussed any of the business scheduled for this meeting. After staying quite late to accomplish it, I told him I was never going to do that again and, furthermore, our next visit would be quite different. He looked a little hurt, apologized and asked what was going to be different. I said I would tell him shortly before coming to Santa Clara.

Just prior to the next visit, I called to arrange the meeting, told him to pack an overnight bag and be at the NSC front door at 3:00 P.M. sharp. He wanted to know where we were going, but all I would tell him was he would be back in the office at ten o'clock the next morning. He was waiting at the appointed hour, and we drove south to the beautiful Quail Lodge in Carmel. After checking in, we found a quiet lobby alcove and settled in for a long talk that resolved all our business issues. During cocktails I suggested this was a far more productive session than the last time we were together. He laughingly agreed, and we went on to have a delightful dinner in a picturesque small French restaurant. Jack got the message, and all visits thereafter had minimal interruption.

At year-end, June 30, 1975, we had sales of $3,364,000, up 42%, and a net after-tax profit of $374,000, up 46%. LED semiconductor volume jumped nearly 70% to $1,414,000. Components and systems were $767,000, up 58%. Because the pipeline for the SX-70 had been filled and inventories were being adjusted, Polaroid sales dropped 60% to $199,000.

Of greater significance, 1975 marked the beginning of our formative thinking on how large thick aspheres for projection television might be made. It was also the year we hired Brian Welham.

# 20 Brian H. Welham

In the spring of 1975, Brian called suggesting that we might talk again. Since our last conversations, no progress had been made in finding a good candidate for the top optical technology position. I could not have been more delighted, and immediately arranged a trip to Rochester for a get-together. Don Keyes joined me because I wanted to demonstrate to Brian that we had a high-quality technical person on board whose material and processing skills would complement his optical expertise. I was also interested in Don's evaluation of the candidate.

We had a long talk at the Welham's home regarding what we were doing technically, what our needs were, and the company's good prospects. His wife Jennifer served a superb gourmet dinner, which I took as a good sign as to our prospects of bringing them to Cincinnati. The major project he was leading at Bausch & Lomb was coming to completion, and its management was starting to lose technology focus as interest shifted to consumer products. While that was happening, our future looked brighter than ever.

Brian was born in Wales. After primary and secondary education he went to work for Vickers Armstrong Aircraft, Ltd., a manufacturer of airplanes in England. While at Vickers, he simultaneously studied aeronautical engineering at Kingston Institute of Technology, receiving a degree in 1957. The program also had a heavy emphasis on mechanical engineering. Shortly after graduation he went to Canada, becoming employed by aircraft maker AVRO. Unfortunately, AVRO failed in 1958, so he secured a position with North American Aviation in Columbus, Ohio. Six months later his new employer had a major contract canceled, leaving him without a job. James Ballmer, an engineering friend at North American, had moved to Bausch & Lomb and encouraged Brian to come there. By this time he had enough of the ups and downs common to military aviation contracting, so he took Ballmer's suggestion and joined him in Rochester.

Bausch & Lomb was one of the premier optical companies in the world. It made consumer ophthalmic products, microscopes, photographic lenses as well as highly sophisticated industrial and military optics. Rochester, with its Kodak headquarters, B&L headquarters, huge Xerox presence and University of Rochester optical program, was generally considered the top optical center in the United States. Brian's new employer had an incredible depth of talent in all areas of the field. The mentoring and training from the specialists they had assembled was exceptional. He was fascinated by optics and took complete advantage of the opportunity leading him to positions of greater responsibility. At night he enrolled in courses at the university taught by some of the greatest names in the field.

When his Bausch & Lomb career came to an end, Brian had received six patents, published several technical papers, been the engineering leader for a number of successful complex programs, and was recognized as a very gifted optical engineer. He joined U.S. Precision Lens August 1, 1975, as vice-president, research and engineering.

From the moment he arrived, it was obvious Brian would be of enormous help. While USPL's lenses at the time did not have the precision found in B&L's more sophisticated systems, we did have the challenge of advancing and finding the quality limits of plastic optic technology. He easily made the transition required to comfortably pursue that objective.

The technical leadership problem that had concerned me so much from the beginning, was now solved beyond our hopes. Brian's mechanical engineering and optical experience were a perfect combination. His understanding of design, tolerances and quality level requirements, plus the ability to communicate internally as well as with customers, proved invaluable. Knowing when something was good enough, when a compromise could be made that didn't diminish performance, and how to explain it clearly are special talents he possesses.

Without Brian Welham, USPL today would be a far different and far smaller company. He has been one of our most unique and significant contributors.

Herman L. Buhlmann founded United States
Watch Crystal Manufacturing Company in 1930.

Henry Buhlmann, Herman's son, managed
the company from 1936 to 1964. Under his
leadership a plastic injection molding machine
was acquired to make plastic watch crystals
and, eventually, plastic optics.

#1 - The original Van Dorn injection molding
machine purchased in 1953.

Examples of molded plastic optics.

*3800 square feet location in Brand Photographers building at 1214-1216 Central Parkway. White line shows area occupied by the company.*

*(These photos were taken October 15, 1970 - the day Roger Howe purchased the company).*

*The unimpressive entrance.*

*Phil Almand in the office that someone said looked like it had been furnished with military surplus furniture from the various wars the Nation fought in the last century.*

*The toolroom where mold inserts were made contained a few well aged machines (above).*

*The crowded molding room that housed four injection molding machines (below).*

*Herman Buhlmann's original glass edge grinding machines.*

*A small room that housed quality control, shipping and a lunch area.*

*Roger Howe on his last day at S.D. Warren Company in February 1970.*

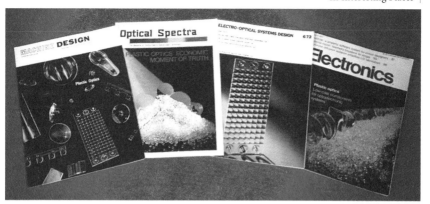

Cover Stories in top engineering magazines on plastic optics were
a major part of the answer to "the marketing problem".

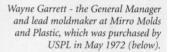

Wayne Garrett - the General Manager
and lead moldmaker at Mirro Molds
and Plastic, which was purchased by
USPL in May 1972 (below).

Bob Campbell, an important early contributor, was
put in charge of production in 1971 (above).

In January 1972 the company moved to a new factory
at 3997 McMann Road.

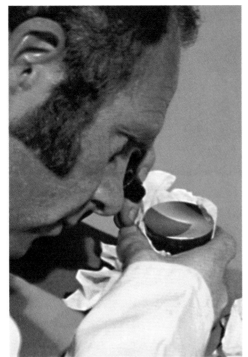

*Digital watch and handheld calculator with LED display boards and bubble lenses.*

*Dick Wessling joined USPL in September 1972 to head mold insert optical grinding and polishing. In the years to come he would do much more.*

*A typical injection mold. The photo shows the two halves each containing four cavities of polished optical inserts along with a finished molding prior to removal of the filling runners.*

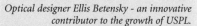

Optical designer Ellis Betensky - an innovative
contributor to the growth of USPL.

Dr. Edwin Land, Polaroid CEO, and his
revolutionary SX-70 camera utilizing three
USPL optical parts - October 1970.

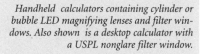

Handheld calculators containing cylinder or
bubble LED magnifying lenses and filter windows. Also shown is a desktop calculator with
a USPL nonglare filter window.

Don Keyes, a molding materials and processing
expert, was employed in October 1974.

Typical injection molding machine.

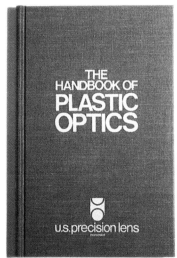

THE HANDBOOK OF
PLASTIC OPTICS.

*Dave Hinchman began his distinguished
career with USPL as Vice President of
Sales and Marketing in September 1973.*

*Brian Welham left a significant
engineering position at Bausch &
Lomb to join USPL in August
1975 as Vice President,
Research & Engineering.*

# Part Two

## *Advancing the Technology*

# 21 A Brief Tutorial on Spherical and Aspherical Lenses

To understand the greatest advantage of plastic optics, one must understand what an aspheric optical surface is and why it is important. Virtually all glass lens elements have spherical surfaces. If a spherical curve is extended, it will eventually meet itself. The rate of curvature change is constant. Designing with only spherical surfaces usually imposes a limit on the designer, which results in the need for additional elements to achieve the desired performance with minimal aberrations.

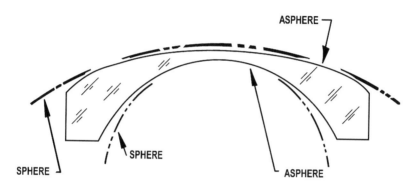

Aspheric lens surfaces do not have these disadvantages. With aspheres, the rate of change in the curve is varied to produce whatever performance characteristics the designer desires—a technical freedom that cannot be overstated in importance. Since the complex aspheric surface needs to only be generated once in the mold, it can be repeatedly reproduced economically by the plastic molding process. Aspheric lens design, moldmaking and measuring require highly-specialized ability and experience. USPL started developing these skills in the early 1970s, and has continually improved upon them to this day. While plastic optics have important cost and configuration advantages, it has been the ability to understand and manufacture precision aspheres that has made our company a truly special maker of optics.

There is no known high-speed manufacturing process to make precision aspheric glass lenses at a reasonable cost because no one has found a way to precisely grind and polish glass blanks quickly at the accuracy levels required.

When Henry Kloss needed an aspheric Schmidt corrector lens to minimize spherical aberrations in his first reflective projection system, he needed a technology from us that was very difficult to find. If the corrector had only required a spherical surface, he might have been able to make it internally or obtained it elsewhere.

# 22 Researching and Developing
## a Breakthrough Manufacturing Technique

We were highly motivated by the belief that if an alternative method to conventional injection molding for making large aspherical lenses could be found, the opportunity to provide optics for projection television would be immense. After the Advent prototype molding failure, Brian Welham, Don Keyes, Dick Wessling and I spent many hours exploring alternative ways to solve the problem. We developed a list of eight different possible techniques. Some were wild in concept, and none had ever been used in mass production. Since hot plastic shrank too much in thick sections while cooling, we tried putting two thin shells together and filling the void with various epoxies and mineral oil. An underwater camera company we supplied had a part that was perfect for these trials. Unfortunately, the epoxy filler cured with big internal air bubbles. We recalled that early black and white TVs often were sold with an additional thin plastic faceplate that was filled with mineral oil to create magnification. We studied mineral oil index of refraction and dispersion characteristics before making another attempt. Those trials also failed as the oil kept leaking onto Brian's desk.

At Combined Optical Industries in 1972, I saw how they made large aspheric magnifiers. The technique was to machine, from a block of plastic, the closest sphere to the final surface desired. The block was then put in a compression molding machine with the correct aspheric curve and reheated. The material near the surface flowed and, with pressure, formed the proper surface. Utilizing molding machine heater bands in a "Rube Goldberg" setup one Saturday, we conducted trials which were successful enough to give us hope, except for one quality characteristic. The plastic surface had hairline stress cracks everywhere that could only be seen with strong magnification. Brian said that would not be acceptable in a multi-element lens. I purchased one of Combined's big magnifiers at an office supply store and discovered they had exactly the same problem. However, for their application, it didn't matter.

Polaroid had been studying single-point diamond machining of metal and plastic. Dick Weeks, the head of Optical Engineering, showed me a beautiful large asphere they had machined in a long cycle on a half-a-million-dollar curve generator. The cost to produce such lenses in volume would have been prohibitive, but the process made an acceptable part.

Others had experienced limited success making small aspheres by a process called replication. A close-fitting glass sphere was put in an aspheric mold interfaced with a thin layer of optical epoxy. When the epoxy cured, the mold was removed from the new aspheric surface. We determined this would be difficult to do in an efficient way with large lenses.

The experiments led us to an idea for a process that has never been disclosed to anyone not under USPL secrecy and non-compete agreements. Its development required two breakthrough ideas that unfortunately must remain secret. The first idea came from Don Keyes. It was a concept that dealt with how the surface could be formed. He found a machinery company where we could run trials. With non-disclosure agreements in place, on January 15, 1976, the two of us were able to successfully demonstrate that piece of the puzzle. Dick Wessling had been working on a technique that would be the final piece. With the first part accomplished, Dick was able to demonstrate a process that would answer the second. We were now able to make aspheric plastic lenses over an inch in thickness and up to eight inches in diameter. In the years to come we would learn to make them even larger.

We had been working intensely on the development as evidenced by my January 6 memo to six key USPL people. Particularly note the second paragraph, which suggests the importance I placed on it.

## INTEROFFICE MEMO

TO:     Dave Hinchman          Dick Wessling

        Bob Campbell           Don Keyes

        Brian Welham           Terry Hood

COPY:   Henry Kloss, President

        Advent Corporation

FROM:   Roger Howe

SUBJECT: Advent Development Program

Today I reviewed, for Henry Kloss, our development work of last weekend and our continuing program for the five-element system. I explained that we were going to be open with him as to our work, but we were doing it on a strictly confidential basis. Obviously, if our approach and techniques are proprietary, it is to both Advent's and our advantage. If we are as successful as we anticipate, there will be a lot of interest in just what it is we are doing.

Because I believe this is one of our most important opportunities and because of the new capital-expenditure implications, I will be involved in the program on a daily basis.

At this point Henry believes we will not need an achromat, so the cementing problem may be eliminated. Also, it appears that all lenses can be made from acrylic. Hopefully, we can still keep the fast system. Edge configuration may now have to be different, so we must have a session and report our thoughts to Henry.

Please review with people reporting to you that our work on larger optics is strictly confidential, and it is likely to remain so even if we get to full production.

The memo falsely implies that Henry knew our process in detail. He had been shown large spherical lenses and did have a general idea of what we were doing, but he was not told any of the key elements.

We purchased the critical pieces of equipment needed to produce optics by what was now called the massive-optics process. On

January 29 Brian and I visited David Gray, the Advent optical designer, to better understand the design he had created for Henry and explore compromises that might make it more manufacturable. Mr. Gray told us it was no problem to design the lens with its many aspheric surfaces, but he didn't think there was any company in the world that could make it. As it turned out there was one—and it was U.S. Precision Lens.

After making successful prototypes, Henry asked us to proceed to production as quickly as possible. We were buying off–the-shelf machines which had to be modified for our process, and building others from scratch. In addition, there was a large amount of tooling that had to be built internally. A small area of the factory was dedicated to the operation. Don Keyes, Brian Welham, Dick Wessling and I were spending virtually all of our time on the project, while Dave Hinchman was tending to Polaroid, the LED makers and other customers who were providing the income to fuel the effort.

It was a very exciting time as we ramped to the late summer introduction of the new big screen TV called the Advent 750. It projected a 72-inch diagonal image that was superior to the earlier model 1000A Schmidt system. Advent promoted its unique four-element, f/1.3 plastic lens system. Actually, there was a fifth plano element for radiation shielding and the lens was in fact closer to f/1.4. It was, however, not uncommon in the optics industry to favorably round f numbers which indicated the amount of light input that would get through the lens. A low f number, or fast lens, was better. For example, an f/1.3 lens would deliver 20% more light to the screen than an f/1.4.

My heavy commitment of time and resources to the new technology throughout the spring and summer was not without a little internal criticism. One of our sales executives widely, but never in my presence, referred to the effort as "Roger's hula hoop," suggesting it was a passing fad. On his own volition, he left us shortly thereafter.

Henry Kloss's three-tube refractive optic approach to making a projection set was not a secret. He had frequent contact with people possessing expertise in the various aspects of television design. One of them was Dr. William Good, of General Electric. Bill Good was one of the inventors of the G.E. light valve projector, which sold for over $50,000. It was a superb device but, due to the cost, was limited to commercial use. He had heard from Henry about our lens development and made an appointment to visit us in June to learn about our capability. It was a meeting that would lead to the creation of the first consumer rear screen projection TV.

Throughout 1976, Advent was in far worse financial condition than we realized. It was a good thing we didn't have knowledge of it early in the year, because the new massive-optic lensmaking development might have been dropped. In the days ahead, Advent would have great difficulty.

Our invention of the massive-optic process to make large aspheric plastic lenses was a fundamentally great idea and would transform the company in ways far beyond anything we could imagine at that time.

# 23 Tumultuous Times with Advent

Later we would learn that back in June of 1975, the State Street Bank in Boston told Henry Kloss either to find new equity money or take it out of their loan. His long-time lawyer, Peter Seamans, contacted Peter Sprague, who was presumed to be a wealthy investor who found it interesting to rescue distressed companies. Sprague was the non-employee chairman of National Semiconductor which, at the time, was probably our largest customer. He acquired control of NSC when it was tiny and had the foresight to bring in Charlie Sporck, an executive from Fairchild Camera, to run it. He also owned interests in a Cambridge specialty retailer called Design Research, and Aston Martin, the British manufacturer of luxury sports cars made famous by James Bond movies.

Sprague invested $575,000 for a 20% interest in Advent and, as part of the deal, received voting control over Henry's stock. It was then doing $17 million in sales, but losing money. Bank debt was $5.3 million. As we were gearing up for production, he hired Pierre Lamond, an operations v.p. from National Semiconductor, to be president. Henry was relegated to a product development role.

Of considerable concern to me were comments from Henry that he was developing a superior, significantly brighter projector utilizing an improved reflective optic system. We had about $12 of content in the 1000A. In the Model 750, our content was $126. If he succeeded with the new reflective approach, it was back to $12 for us. We had made substantial capital investment and created a special manufacturing group that was now in danger of being obsolete almost before there was a full-scale startup.

We needed a backup plan. Brian asked Ellis Betensky to look at Gray's optical design to see how it could be improved upon, particularly with respect to lowering the f number to increase light transmission and, therefore, picture brightness. This was the kind of challenge that Ellis thrived on.

Henry and Peter Sprague clashed over a lot of things. Peter wanted to build a one-piece projector with the picture reflected off a mirror to an attached screen. Henry thought it would drive up cost too much. They had different views on distribution. Finally, on October 21, only weeks after the introduction of the Model 750, Henry resigned. He called me that day to explain what was going on and why he felt it was the right thing to do.

Four days later I wrote to thank him for all he had done for us and to offer our best wishes for the future. The third paragraph of that letter read as follows:

> The real purpose of this letter is to tell you that as a result of your willingness to look for the new optical approach, and for your confidence in U.S. Precision Lens, our company may well have developed an exciting new business that will ultimately be bigger than our existing activity. Our company, too, speculated on the development, but strictly because of you personally. What you did was give us enough inspiration, understanding and latitude to pursue a challenge that, under almost any other conditions, we would have declined. For me, it was not an Advent bet but rather a Henry Kloss bet. At this point we are much like Advent. If we fail with what we have now, it is clearly our mistake and no one else's.

A year later, Jacqueline Kloss told me that letter meant a great deal, both to Henry and her, because it was the only one they received at what was then a very low point. Henry would be back, and we would always be good friends.

Early in 1977, Pierre Lamond's number-two man, Ed Cobb, asked me to come to Boston to meet with them on an important matter. I met with them soon thereafter. Cobb said we would have to significantly lower our prices or they would move the business to a new competitor. I had no idea who the competitor might be; however, we always assumed others would get into the business. I resisted, explaining that we were not close to recovering our investment and that our prices were fair. Lamond then quietly said to Cobb, "Well, I guess that's it—let's get going with Bell & Howell." This was most interesting because at that very same time Bell & Howell was approaching us for similar lenses to be used in projection systems. Knowing that, I got up and said, "Gentlemen, we are deeply indebted to Advent because you brought us into this business. We understand that you have to do what is right for your company. You should know that we will cooperate in the transition in every way, but we would like to get it over with as quickly as possible because we have a lot of new demand for the capacity." As I was about to leave, Cobb said, "Now let's not be hasty—we don't really want to move away from U.S. Precision Lens." Their crude bluff didn't work.

In May of 1977, Sprague fired Lamond and personally took over as president. Sales took a big jump to $36 million as a result of the Model 750 introduction, and the company actually had a net profit of $2.3 million. Manufacturing, in a very expensive and disruptive move, was relocated to Portsmouth, New Hampshire. In 1976 Advent sold 5,300 sets and in 1977 would more than double that number.

From the time of his departure, Henry had been feverishly developing the new reflective system which he expected would first be sold to Advent. We had Ellis working on an f/1.0 design that would be a vastly superior alternative to the Gray creation. At f/1.0, ours would have nearly 100% more picture brightness. By March of 1978, Henry had successfully demonstrated his improved reflective projection tube, and we had Ellis's new design. Both were offered to Advent and, astonishingly, both were turned down. Henry decided to start a new projection TV company called Kloss Video, and we prototyped the f/1.0 lens so it could be demonstrated to others. It would take a few years, but that was the beginning of the end for Advent.

# 24 The First Rear Screen Projection TV

After Dr. Good visited us in late June of 1976, General Electric started working on a concept for a different kind of projection device. Unlike all previous versions, the television tube and lens that projected the image would be out of sight behind a translucent screen. Its appearance would be more like a normal TV.

The G.E. television business was headquartered in Portsmouth, Virginia and managed by Fred Wellner, a tough former World War II German submarine commander. Reporting to him was the Syracuse picture tube division, which also had responsibility for the expensive light valve projection products. Syracuse was managed by Dr. Gary Carlson, a bright innovator who was recruited from Texas Instruments.

Carlson had his people working on various ways to make a reasonably-priced consumer version of the light valve, but cost made it a hopeless task. Due to all the market attention Advent was getting, he had them shift their effort to a more conventional television tube solution. His concept was to build a 13-inch tube that would be far brighter than what was regularly being marketed. Its image would be piped through a large lens and bounced off an internal mirror to the back side of the screen. It was a terrific idea, but Wellner was not enthusiastic. Carlson pressed forward anyway but was under clear instructions not to show the device to any higher-up executives.

One day he was taking David Dance, a G.E. vice-chairman, for a tour of the factory and laboratories. As they were walking down a hall, someone popped out of a door and Dance looked in, exclaiming, "What's that?" Carlson had no choice but to take him in to see the rear screen consumer TV mockup. He thought it was fantastic and encouraged everyone to make the project a priority. He also raved about it to Wellner, who was now furious with Carlson. In spite of an excellent explanation, Wellner had a hard time forgiving Gary. I have often wondered if there ever would have been a rear screen consumer projection TV if someone hadn't opened the door to that lab the day Dave Dance was taking a tour of the Syracuse facilities.

A G.E. team visited Cincinnati to determine if USPL could use the massive-optics process to make a large multi-element lens for the 13-inch souped-up tube. They needed a design, so we had Ellis go to work on it. Brian and I soon were visiting Syracuse regularly, and they were frequently bringing various players to Cincinnati. The optical design called for three thick aspheric elements 6.84-inches in diameter. Its speed was f/1.6. Over the next year we continue prototyping and making improvements. They wanted the lenses coated to eliminate reflections. Each surface reflects about 4% of the light hitting it; so with six surfaces, 24% of the light would be lost if they did not have an anti-reflective coating.

Sophisticated optical coatings were deposited by evaporating special metals in a vacuum chamber that held the lenses. It was a new technology for us, and we were reluctant to make the large expenditure that would be required. Brian sent prototype elements to Bausch & Lomb to be coated, and to determine if they could do the work for us. Although the coating was excellent, the quotes were quite high. Also, we worried about the elements being scratched from too much handling if they were sent out. Finally, at G.E.'s insistence, we made the decision to acquire internal capability. We hired Bob Schaeffer, the talented owner of Evaporated Coatings in Philadelphia, to provide advice on what equipment to buy and to train our people. He did an excellent job of getting us in the business and would help us again as we upgraded to more complex multilayer coatings in the future.

G.E. was very worried about being so dependent on such a small company as USPL. They demanded to see the massive-optics process, which we considered such an important secret. The issue escalated to a serious impasse, and something had to be done to break our frozen positions. Their concern was that after making the huge investment to build the set, if we didn't perform they would have no way to continue producing it. I suggested a solution. It called for us to document every aspect of the process on microfilm and have Arthur Andersen verify that it had been done. If for any reason USPL did not meet their requirements, we would give them the documentation. They finally agreed, but it really annoyed them that we would not reveal how the lenses were made.

The new 45-inch rear screen set was to be introduced in April of 1978. In March, I was asked to come to Syracuse to meet with Fred Wellner. After a stiff introduction, he said, "Young man, do you know how important your company's performance is to General Electric? Will you be there for us?" I said, "Yes sir, Mr. Wellner." And then I added, "Mr. Wellner, we have bet a large part of our company on the success of this product. In case of unforeseen events, will you be there for us?" He said, "Of course we will—how could you possibly question General Electric on such a matter?" Within a year, we would learn just how important my question was.

On April 12, 1978, G.E. introduced the first rear screen projection TV with much fanfare at a press conference in Portsmouth. Shortly before it started, Wellner's boss, David Dance, asked about the retail price. He was told $2,495. Next Dance asked how many they could sell, to which Wellner replied, "All we can make." Dance then told him to raise the price to $2,800. The introduction was big news in the TV industry and was widely reported in the press. The G.E. TV marketing manager forecast sales of 125,000 units in 1978, climbing to 500,000 by 1983. We were being warned that orders might be significantly increased.

Television makers throughout the world watched with interest what Kloss had done at Advent, but they really took notice when the prestigious General Electric Corporation entered the projection business. Virtually every large consumer electronics company started a feasibility program to determine if they should make a projection set.

The G.E. product utilized an internally-designed innovative Fresnel screen made by the Optical Sciences Group in California, which directed the image to the viewing area, thereby significantly increasing brightness. This meant one had to be directly in front of the screen to see the picture because it dimmed and then disappeared with fairly short viewer movement to the left or right. Even when one was directly in the viewing area, it was deficient in brightness if there was high ambient light in the room.

Production scheduling was a nightmare. Orders would be increased, cut back, and then increased again. Part of the problem was that the Optical Sciences Group was having serious difficulty making the screen. Nevertheless, for over a year it represented good business for us. No new single tube model was ever introduced, as G.E. realized they needed more brightness and better viewing angle than they could get with this first design.

Near the end of the product's life cycle, we received a huge order cancellation from the Portsmouth purchasing department. A big shipment of our $135 lenses was ready to go and they didn't want them. Beyond that, we had hundreds of thousands of dollars in raw material and work in-process that was covered by purchase orders. The prospect of cancellations was intolerable; we were in shock.

I went to Portsmouth to see the buyer, a fellow we had been working with from the beginning. During an unusually long wait in the lobby, I picked up and read an excellent booklet on G.E. values and ethics. It was in complete conflict with what they were trying to do to us. The buyer took a firm stance, insisting on the cancellation, and was offering us nothing in spite of the fact that we had firm purchase orders. In the extremely heated session, I finally pulled the ethics booklet out of my pocket, held it close to him, and asked if it had any meaning at all or was it just a bunch of nice words. He didn't want to discuss it and he didn't want the lenses. I was furious, got up, told him I was leaving his office to go to General Electric's headquarters in

Fairfield, Connecticut, where I would sit in the lobby for as long as it took to see chairman and CEO, Reginald Jones, because I wanted to ask him if the booklet had any meaning. Now the buyer was in shock. He said, "You wouldn't do that, would you?" My reply was, "Why not? This is a gigantic loss for us, and we have nothing to lose." Then I told him about Fred Wellner asking, "Will you be there for us?" my asking Wellner the same question, and his indignant response of how could I question G.E. on such a matter. Now convinced I would do it, the buyer mellowed quickly saying we had to work it out, and he didn't want me going to Fairfield. A solution was finally reached that we could live with, but it was a most unpleasant experience.

General Electric is one of the best companies in the world. It is not unlike other high-achieving firms with strong ethical principles as part of their culture. Sometimes, however, when intense pressure for performance is pushed down to the lower levels, individuals do things that people at the top would not approve. This was one of those cases. It is not smart to do what I did unless there is no alternative, as it was in this instance. I fully expected the buyer would somehow make us pay for forcing him to back down, but he did not. Our relationship with G.E. was to remain excellent, and they would introduce many new models in the years ahead using our lenses.

# 25 1976

As moldmaking and molding operations continued to grow, it was apparent by the end of 1975 that we needed to strengthen our manufacturing organization. Bob Campbell was stretched quite thin in overseeing both functions, so upgrading his key support positions became an objective. The first job to be filled was manager of plastic molding. We were looking for someone with injection molding management experience to oversee the plastic molding department. Our search firm had difficulty finding candidates locally that were acceptable; so, in collaboration with other firms, they broadened the recruiting geographically. The most qualified candidate was found in Massachusetts, working for Texas Instruments at their Attleboro manufacturing facilities. His name was Peter F. Cutler, an electrical engineer who had graduated from Tufts University. His position and experience at T.I. gave him the exact qualifications we were seeking. In November 1975 he visited, and after a brief negotiation agreed to join us.

Pete and his wife, Sally, moved to Cincinnati for a start date of February 2, 1976.

During our negotiation I learned that he didn't like snow. He asked if he should bring his snow blower. To the best of my knowledge, it is the only time I ever consciously didn't tell the truth. I told him it didn't snow in Cincinnati, and that he should sell it. That cinched our agreement. Since that first winter, there has never been a time we have been with the Cutlers, that Sally hasn't reminded me of it.

Pete did a superb job of running our molding operations and overseeing its phenomenal growth. In 1981, he was made operations manager for the standard optics division, which encompassed all manufacturing areas other than massive-optics. He was elected vice-president of optical components in April 1986. The standard optics designation was changed to optical components. In that position he was also responsible for making the extremely important housings for television lenses. Until his retirement in 1998 he was an important contributor to the building of the company.

We had become cramped for space everywhere. To alleviate the problem, an 18,000 square-foot addition was put on the back of the plant, and office additions totaling 8,000 square feet were put on both sides of the front section. The office part involved a complete architectural elevation update. This total build of 26,000 square feet doubled our space, taking the pressure off for awhile. Whenever possible we purchased additional land, eventually increasing holdings from the original two acres to ten acres on the west side of McMann Road.

The financing demands for expansion of the massive-optics process were uncertain because there was no basis on which to forecast growth. We suspected it could be substantial, so a $1.5 million revolver loan agreement was established at the First National Bank (now Firstar Corporation). The outstanding balance at fiscal year end was $160,000.

The massive-optics process was being improved as we researched its various facets and made discoveries. Standard machinery modified to our specifications was no longer acceptable. Don Keyes was building development machines that advanced the technology, but the production outlook required that much more be done. In yet another example of how one event leads to another, Brian Welham had become reacquainted with his old friend, Jim Ballmer. Ballmer was the person whom Brian first knew at North American Aviation and would later encourage him to come to Bausch & Lomb. Jim had left B&L to become the chief engineer for the McGregor Sporting Goods Company in Cincinnati. Now he was working for a firm that provided contract engineers to companies on a temporary basis. Brian brought him in to help us with the machine-building problem.

Jim Ballmer was a unique and highly talented mechanical engineer who had a wide range of experiences, including the building of equipment to produce optics. He quickly designed reasonably complex, yet cost-effective machines for our processes and oversaw their construction. Because of Jim's work, we were able to ramp our production capacity in advance of demand. After becoming a full-time associate, in the years ahead he would find many ways to improve the massive-optics process. Furthermore, he would be called on to lead or give advice on all new major engineering projects. Jim was a great contributor to the success of USPL before his retirement in 1990.

When fiscal year 1976 ended on June 30, sales were $4,461,000, up 33%, and net after-tax profits were $428,000, up 14%. Because all massive-optic process development costs were written off as incurred, profits did not show a corresponding increase to sales.

Polaroid sales had significantly rebounded to $776,000, up 290%, due to the addition of a low-cost version of the SX-70 called the Pronto, for which we also made viewfinder lenses. Components and systems continued to grow nicely, up 36%, breaking the $1,000,000 level for the first time. LED lens volume was down 18%, to $1,158,000, as manufacturers adjusted production to correct for the previous year's over building.

On September 2, Dave Hinchman was elected to the board of directors, and made president and chief operating officer. In addition to recognizing his excellent performance, the primary reason for the change was to further indicate his important role to customers and enhance his position internally. I was spending the majority of my time on the technology, people, sales and capacity issues of the projection television business. Dave filled the void caused by my partially backing away from the LED, Polaroid and components activities.

Ellis Betensky was designing innovative new zoom photographic lenses for Vivitar. He was excited about the improvements that could be gained by substituting plastic aspheres for some of the glass elements. He was also doing design work for the Joseph Schneider Company, a world-renowned glass optics maker in Bad

Kreuznach, Germany. To gain our interest in working with him on Vivitar developments and on a project he was doing for Schneider, he suggested I attend Photokina, the world's largest photographic show in Cologne, Germany. It was an interesting experience that gave me a chance to meet all of Vivitar's senior people, and marked the beginning of a relationship that would eventually lead to one of our most significant strategic moves.

At Schneider, we received orders for two photographic attachment lenses that could only be made by our massive-optics process. With acquaintances having been made and our credibility established, this visit laid the groundwork for our receiving a very large projection TV lens order in a little over a year.

Throughout the year, from time to time I would observe that Bob Campbell was becoming what is commonly called "burned out." He had tirelessly worked to help create the enterprise, which was now twenty-four times bigger than when we met. Bob was one of the key people to be given an opportunity to buy the company's shares and, because of an early purchase, had a substantial paper profit. Our policy was that shares could be bought at book value, which was far below real value. Any shareholder who left the company was required to sell shares back at the present book value. Because of his USPL holdings, he had enough money to start a small general tooling and moldmaking company, a desire of his we had talked about from time to time.

In a communication to our associates, I made the following comments.

> This announcement is offered with mixed emotions because for me, it has a happy and unhappy side to it. A few years ago, Bob Campbell told me that someday he would like to start a small business of his own. He has again expressed this desire and, as a result, will be leaving us October 15 to start a tooling company. Bob will actively do consulting work for us until December 1, and after that will help us in any way he can. I

am happy for Bob in that he really will be doing what he wants to do and can look forward to exciting personal times. On the other side of the coin, I am saddened because Bob has been one of the most able and loyal builders of USPL. For this I will always be grateful.

The good news is that when Bob gets his operation going, we expect him to become one of our tooling suppliers.

The help Bob Campbell gave me at the beginning was critical to our success. This story might have been far different if he had not been there on October 15, 1970.

With Pete Cutler running the molding operation, there was no problem in that area due to Bob's departure. Management of mold-making, however, was an issue. Fortunately, we were able to recruit an experienced person from General Motors in Dayton. Dick Miller joined us to lead that important function, reporting to Brian Welham.

# 26 The One-eyed Monsters

In early 1976, we began to see a new type of projection TV being pro-moted. Small entrepreneurs who called themselves projection TV manufacturers would buy a 13-inch TV set, a projection lens and a screen. They switched a couple of wires to invert the image so when projected through a lens it would be proper. The two functions per-formed by these so-called makers was to build a simple shroud that covered the picture tube and had the lens, and to do the selling. The preferred 13-inch TV was a Sony because it had superior brightness. The first glass lenses used were f/2.8 with three or four elements. The screens were usually 50-inch.

The projected image from some of these single lens systems was surprisingly good, but the brightness was woefully inadequate. To compensate for that deficiency, the sets had to be viewed under conditions of little or no ambient room light. Because the most prominent feature one noticed was the lens mounted in an often crudely-designed shroud, someone called them one-eyed monsters. It was a name that stuck.

Late in the year, we received a few inquiries for lenses that could be used in these devices. We did not make a lens that could be used for the application, but we were thinking about it. Ellis was put to work to design a cost-effective f/1.9 lens that would provide twice the light output of the f/2.8 glass lenses that were being used. The second design iteration was successful. It contained three 6.1-inch elements mounted in a black clamshell mount. We called it the Beta II. Its 100% jump in light output made it vastly superior to the f/2.8s, but still deficient from what was needed for good viewing in medium ambient light. By late spring of 1977, samples were being sent to a few prospects.

The Beta II business exposed us to an incredible array of characters that were far different from anyone we had ever done business with before. Most were operating with hardly any resources, but to hear them talk, it sounded as if they were going to take over the world. When visiting, they frequently had two questions: "Is your company for sale?" and "Can you keep up with us?" It was difficult to keep a straight face and not offend them. Many found us, and from advertisements and trade publication lists, we found the others.

In June, we booked a suite at the Blackstone Hotel in Chicago during the Consumer Electronics Show. Invitations were sent to all our prospects to come for demonstrations of the new Beta II single-tube projection lens. The word quickly spread that our lens was an important development, and many so-called manufacturers came to take a look.

Don Hummel, a young sales engineer, was responsible for most of the prospects and customers. I was involved with just three of the larger ones. Don divided the thirty to forty makers into three categories: the "A list," "also-rans," and "cats and dogs." The "A list" had names like View Point, Projector Beam, Magna Vision and Futurevision. The others also sounded important. Our prices ranged from $135 in quantities of one to twenty-four to $92 at the 100 to 500 level. We also sold an optional mounting flange for $8 to $10. With these buyers, terms were cash in advance or C.O.D. with certified check or irrevocable letter of credit. No one ever used the letter of credit, and no one could understand why we wouldn't sell on an open account with thirty-day terms.

A number wanted us to sell the Beta to them on an exclusive basis, since they were promising to buy such large quantities. The requests and rumors were so common that in a widely distributed price list I added the following paragraph:

> We would like to put to rest a rather persistent rumor that certain companies may become exclusive buyers of the Beta. There are no exclusives now, and there will be none in the future. We carry

substantial inventory for immediate shipment and have the manufacturing capacity for much higher volume.

Most of the buyers purchased 50 to 100 at a time, with only three taking larger quantities. From July 1977 through June 1978, our 1978 fiscal year, we sold nearly 10,000 Beta's, resulting in over $1 million of volume. It accounted for 14% of the year's sales, so it was an important product.

The most promising of these makers was a Lakeland, Florida startup called View Point. It was a subsidiary of a reasonably-successful small insurance company. After establishing that USPL was not for sale and receiving our assurances that we could keep up with their demand, they asked us to ship everything we could make as quickly as possible. In two months we sent them 2,500 lenses, which caused them to call and begging us to stop. They were better at ordering components than selling TVs. In spite of the insurance company backing, we demanded and received cash in advance. It was a good policy, as the venture collapsed shortly thereafter.

In Los Angeles there was a flamboyant British fellow by the name of John Bloom whose company, Projector Beam, aggressively advertised its unit in Southern California. Bloom had once made a fortune and a big name for himself selling low cost washing machines in England. For a couple of years he was the country's best known entrepreneur, as evidenced by the many magazine features he showed us. The washing machine company had a spectacular collapse, leaving John without his plane and yacht. There were so many people pursuing him because of broken commitments and debts, he fled to the United States. His wife controlled what little was left of their money with an iron fist, as he came up with ideas to make the next fortune. After running a semi-successful theme restaurant, he moved on to projection television.

We had sold 100 lenses to Projector Beam in December of 1977. John called to explain that he was going to build a very large TV business, and that we should visit him as quickly as possible in Los Angeles. I was given details about a Chicago financial backer and urged to call him for reference. The Chicago backer was a big electrical contractor who was helping to build towns in Saudi Arabia. He had only met Bloom on a couple of occasions and was spending most of his time out of the country. He did confirm that he was a substantial investor.

We were going to the Las Vegas Consumer Electronics Show in January anyway, so I told Bloom we would come out to see him. He said he would have his Chicago friend's jet pick us up in Cincinnati and added "By the way, bring along fifty Betas because I need them right away." Brian Welham, Don Hummel and I loaded the lenses into the Aero Commander once owned by Elvis Presley and took our first ride on a private jet. As the plane came to a stop at Los Angeles International Airport, there was John Bloom pulling up in a gold Rolls Royce followed by a panel truck. The lenses were hustled into the truck, and we were instructed to make ourselves comfortable in the Rolls for the trip to our hotel in Beverly Hills. Although Projector Beam's terms were cash in advance, after getting out of the plane he sent for us and, before getting in the Rolls, I didn't have the nerve to demand payment before he took the lenses.

The Blooms entertained us in their Beverly Hills home and at one of the chic clubs for dinner. While we had a good time, it wasn't a trip that gave us confidence in the future of Projector Beam. April was the last month we shipped to this company. In our brief association, the flashy Mr. Bloom bought about $75,000 worth of lenses from us and, of course, did not pay for the fifty we delivered on the jet. The last I heard of him was when *Forbes* magazine did a story on people who could not go back to their country because of legal problems. He had moved to an island off the coast of Spain and was shown in a full-page photo as one who had distinguished himself by not being able to go back to England or the United States.

The most flamboyant of them all was Californian Earl "Mad Man" Muntz. In the 1950s and 1960s, he was a high-profile entrepreneur who had been involved in a number of businesses. For a couple of years he was a frequent guest on television's *Tonight Show*. Now he was making single-tube, one-piece projection TVs that had a pop-out mirror which reflected the image to the screen. I had heard about Muntz for years, primarily because of his black-and-white TV company that, at one time, had been a heavy advertiser.

We heard that he was selling large numbers of sets, so I called to tell him about the Beta. He told me to come out and see him. His small factory in Van Nuys was attached to a fair sized retail store. After giving him a Beta and watching it be successfully tested late in the afternoon he said, "Let's go: you're staying at my place tonight." As we made the eight-minute drive to his Encino home, he pointed out a large, beautiful home he once owned, along with a couple of celebrity stories about things that happened there. Earl lived in a modest but nice home. Like the outside and inside of his car, nearly everything in the house was white. This included the walls, furniture, carpeting and drapes. At his bar, he regaled me with stories of his past. As a kid in Chicago in the 1930s, he converted table-top radios for use in cars. Just as World War II started, he moved to Glendale and started selling used cars. It was then a young advertising friend gave him the name "Mad Man Muntz," promoting the idea that he made crazy auto deals favoring the customers. He dressed up in a Napoleon Bonaparte uniform for the ads and when working the car lot.

Prior to the visit, I was telling a friend that I was going to see "Mad Man Muntz." He told me that while driving down Sunset Boulevard one day in the 1940s, Muntz, dressed in his Napoleon outfit, jumped off the curb in front of his car waving a fist full of bills and yelling, "I'll buy that car for cash."

I asked Earl to tell me his life story, where he made the most money, and what was his biggest mistake. After the used car business during the mid-1940s he set up all distribution for the Kaiser Automobile Company, which became quite large after the war. Then he started the Muntz TV Company. His sets worked fairly well in metropolitan areas, but once one traveled some distance from the transmitter, the picture faded. The joke was that Earl bought an RCA set and started cutting wires out of it. When it stopped working, he reattached the last one and copied what was left. Muntz TV failed in the early 1950s, so he started a new company and became one of the two big promoters of eight-track stereo systems for cars. When that failed, he moved into projection television. He said the most money he ever made was setting up the Kaiser distribution system, and the biggest mistake was with the eight-track tape business. He made a deal in the early 1970s to sell the company to Gulf Western Industries for $17 million worth of Gulf stock and $500,000 in cash. When he went to the closing, he slammed his fist down on the table and said, "I won't do it for a dime less than $750,000 in cash." After a long pause he slowly walked out, expecting someone to say, "O.K., come on back Earl, we'll meet your terms." They didn't! A month later the eight-track business collapsed and his company went into bankruptcy. I had to agree with him that it seemed like a fairly big mistake.

His bar was at one end of the living room, which contained three 50-inch projection TVs, one regular set and a small monitor. Two of the walls were covered with mirrors. As we sat at his bar, after at least a couple of drinks, this sixty-two-year-old man decided to talk about our lens prices. After slipping a tape into the VCR and turning it on, he proceeded to say, "Roger, you don't know anything about pricing your products." Then he advised me on all the things we should be doing differently which, of course, would mean lower prices for him. The three large and two small TVs were prominently displaying the content of the tape. It was also being reflected from the two mirrored walls. The tape was a well-known infamous X-rated movie that was being highly criticized at the time. Finally I said, "Earl, I know you are trying to teach me how to run my company, but I find it very difficult to discuss business under these conditions." He laughed, gave up, turned the tape off, and we had a nice dinner with his attractive late-thirtyish girlfriend, a former Miss Kansas City who, of course, was dressed in white.

"Mad Man Muntz" was such an unbelievable character that I took Dave out to see him. As with my first visit, we were invited to his home for cocktails. Now my earlier stories had credibility. Earl had recently seen someone hypnotized, so that night at dinner he decided to hypnotize me. It went on and on with nothing happening, so I finally pretended to be under his spell. He was delighted.

Muntz was one of the largest Beta lens customers, and the only one we ever sold on an open account. He always paid his bills.

The single-tube projectors were quite inferior to the three-tube versions that were growing in popularity. Yet, to our total amazement, Beta lenses were a significant part of our sales into the early 1980s. Twenty years after its introduction, USPL still receives an occasional request for them.

# 27 Auto-Focus Cameras

In 1964, Norman Stauffer, a Honeywell scientist, invented a successful automatic focusing system for slide projectors. For the next ten years, he worked with a small group of engineers to create such a device for 35mm cameras. In 1975 he succeeded, and Honeywell signed licensing agreements with thirteen makers. Prior to this development all variable-focus cameras had to be adjusted manually, requiring the user to estimate the distance to the object being photographed. The Stauffer auto-focus system would automate the function and make it more precise.

Early in 1977, Honeywell asked if we would be interested in making two lenses that would be part of the module including their electronic logic chip, a mirrored prism and a housing. We not only bid on the lens but also, proposed that the entire assembly be done at USPL. This meant receiving chips, housings and prisms from other companies, to which our lenses would be added on an assembly line we would be responsible for building.

For years, we thought optical component assembly was a value-added activity that could be a logical extension of our service. It made sense because we understood optical function and testing. In spite of this, no significant assembly business had ever been found. The Honeywell project would become the first.

Mark Tausch, a young engineer, was given the job of building the assembly system. He made a large table with a chain carrier in an embedded track that carried the housing to each work station where parts were added, gluing was performed and the unit was tested. Operators behind safety guards fed the line. It was capable of producing 720 per hour of the first generation VM01 module. Our price for two lenses and the assembly was in excess of $1.00.

The system worked through a process where the photo target image was received in two small windows on the front of the camera, then directed to the integrated circuit sensor chip. Its output was connected to a control device that mechanically moved the lens into focus just before the shutter was opened. It worked beautifully. Honeywell's first customer was Japanese camera maker Konica, which put the device in its compact model C35.

It was promoted as a "dream camera" that took the guesswork out of photography through automation. After introduction in the spring of 1978, it immediately became a big success, causing nearly all the other makers to go to Honeywell for modules. It was clear 35mm cameras would never be the same. That fiscal year our sales of the module exceeded $500,000, accounting for slightly over 10% of our volume.

A second-generation module was designed for single-lens reflex cameras utilizing one of the most interesting injection-molded plastic lenses we ever made. Its minute finished size was much smaller than a clip on a pen and contained fifty-two precise lenses in two rows that were so tiny they could only be seen under high magnification. The optics were used in conjunction with a CCD (charged couple device) that fed the information to the chip, causing it to move the lens to proper focus.

From fiscal 1977 through 1988, we made many millions of auto-focus optics for Honeywell, bringing in sales of $12.6 million. The program finally died because camera makers started making their own systems, which led to an intense patent challenge by Honeywell. In the mid-1990s, the courts held there indeed had been infringement, and Honeywell was given huge awards from all the manufacturers which were in violation. It is impossible to know how many modules we would have sold over the years if the judgement had been made shortly after the infringement began. Nevertheless, it was a wonderful business in the eleven years that we worked and had such a fine relationship with Honeywell.

# 28 The Business Philosophy of U.S. Precision Lens

In 1977, I observed that we had many new associates who really didn't know much about our values and culture. Those who had been around for awhile generally understood our philosophy though little of it had been documented. It seemed to me we should be as explicit as possible on the subject, so I wrote the following and had Dave, as well as others, make suggestions that were incorporated into it. We had it printed as a small booklet and distributed it at associate meetings where the points were reviewed and discussed in detail. Although it wasn't planned for use with customers, a number of them were impressed by the statement and asked for copies.

Looking at it over twenty years later, there isn't anything substantive about it I would change. It read as follows:

Every company has a philosophy. It is a set of beliefs on which it premises all its policies and actions. Another definition could be "the way we do things around here." Although company philosophies are generally not put into writing, we have attempted to do so. Understanding and accepting the U.S. Precision Lens philosophy is important to the success of our company and its people.

I. **Regarding Ethics**

We maintain high ethical standards with customers, employees, suppliers, the Government and everyone else who comes in contact with us. We are proud to be known as an honest company.

II. **Regarding Our Business**

1. *Our Specialty*

We are specialists in manufacturing optical elements made of plastic. We do not involve ourselves in anything outside of our specialty. Whatever we do must relate to an optical function.

2. *The Technology*

We are striving to be the best at what we do. We must not only keep up with our technology, but be a positive force and leader in furthering it.

3. *Profit*

Our objective is to earn a fair profit. We believe that a fair profit is vital to the well-being of our people, the well-being of our customers, and to the furthering of our technology.

III. **Regarding Our Customers**

1. *The Partnership*

Customers who consider employing us are looking for a partner. They want us to be an extension of their efforts to build something for which we supply only one or a few of many assembled components. They are counting on us to meet time, quality and quantity commitments.

2. *Anticipating Problems*

Frequently our customers have little knowledge of plastic optics. It is our duty to help them to anticipate and avoid potential problems.

3. *Ability to Supply Obligation*

Because of our specialty and high tooling investment, we frequently are a sole source of supply. As such, we must address ourselves to every downside risk that may concern a customer.

4. *Our Compensation*

We expect our customers to pay us according to our terms and conditions of sale.

IV. **Regarding Personal Performance**

1. *Adjusting to Change*

Our personal and corporate success requires that we not only adjust to change, but that we also help to bring it about. The world we live in, our economy, our customers and our technology are all changing rapidly. To be effective, we must constructively deal with it.

2. *Decision-Making Based on Facts*

Sound decision-making requires that we understand what we are looking for and get as many facts as practical on issues before taking positions. Anticipating problems is a critical part of our decision-making.

3. *Growth*

As our business grows, we as individuals must grow with it. We must become more qualified in handling the increasing demands of our jobs. We must continually evaluate ourselves and actively seek ways to improve our skills and performance. The company is committed to providing opportunities for growth.

4. *Communicating*

Successful communication with one another, our customers and suppliers is of vital importance. We must accept that communication becomes more difficult with company growth and, therefore, requires greater effort on everyone's part. It is the communicator's responsibility to be sure he or she is heard and clearly understood.

5. *Being a Team Player*

The products of our company require that many different skills be brought together. How well we interact is significant in determining our success. Because every area of our business depends on all the others, we want team players with a positive attitude.

6. *Going Out of Our Way to Help the New Employee*

Every new employee has to deal with many unfamiliarities and, therefore, has apprehension about facing the problems of a new job. We expect our people to welcome new employees and help them learn who does what, how we do things and anything else that quickly makes them comfortable contributors.

V. **U.S. Precision Lens as a Place to Work**

1. *Fairness*

   Our objective is to have rules, policies and procedures that are reasonable and fair so that employees can expect decent treatment regardless of the situation.

2. *Dignity and Respect*

   It is the responsibility of everyone to treat all fellow employees with dignity and respect.

3. *An Enjoyable Climate*

   We strive to make our company a pleasant place to work. All of us share the responsibility for creating a warm and friendly atmosphere.

4. *Pay and Benefits*

   In relation to similar industry in the geographic areas in which we operate, we should be progressive in our wage and benefit policies.

5. *Safety and Working Conditions*

   All employees have responsibility for keeping our working conditions safe, comfortable and clean.

6. *We are an Equal Opportunity Employer*

   We do not tolerate any form of race, religion, nationality or sex discrimination.

# 29 1977

The activity that Phil Almand enjoyed the most was talking to customers about unusual applications for plastic optics. Unfortunately, most of these potential projects were in the early stages of development and had low volume projections. Because of this, he had gravitated to a position of only having minor involvement in our mainstream business sectors. Phil was frustrated by his role not resulting in substantial new business, and some of our associates were frustrated by Phil's lack of productivity. On more than one occasion I had to point out to the critics that Phil had played a vital role in helping me understand the business in the earliest days and because of that, we would always have a place for him. The situation, however, was somewhat of a problem for Phil and the company.

One day in February Dave called me into his office, where Phil announced he was resigning with the objective of starting a custom molding business. We had been through much together since I bought the business from his father-in-law in 1970. I expressed my gratitude for all that he had done and offered our best wishes for his future. The second of the two key players at the beginning of my ownership had now departed to attempt to build his own business.

Because we were still uncertain about the future of the massive-optics television business and because of a lack of space, we leased an area from a neighboring company to make the large lenses for Advent. It wasn't long before it became apparent we were going to grow out of it. So the two molding rooms were consolidated into one, allowing the second to become massive-optics manufacturing. This meant selling eight of the older marginally-useful molding machines and putting the remaining best twenty-four together.

The 1977 fiscal year ending June 30, was financially disappointing, and yet encouraging in other ways. Order levels widely fluctuated, causing the production mix to be inefficient. In February we had 320 associates; by the end of March the number had dropped to 260 as it was necessary to bring employment into line with demand. Later, it was increased again. Moldmaking sales were down substantially from the year before, but massive-optics had gone from insignificant volume to nearly $1,200,000 primarily for Advent. Total sales were up a meager 7% to $4,770,000, and net after-tax profits were down 65% to only $148,000. Long-term debt climbed to $675,000, up over $400,000, and fixed assets grew over $500,000. The heavy massive-optic development expenses and unfavorable production mix had taken a toll; however, the stage was being set for a great fiscal 1978.

Calculator makers continued to suffer from brutal competition although our sales grew 39% in the year that had just ended. Litronix had been struggling to keep up with National Semiconductor and Texas Instruments, but couldn't. It was sold to the German electronics giant, Siemens. Our business with them continued, but on a declining path as the company was repositioned to take advantage of other markets.

In May, Brian and I visited Schneider in Germany, who had become a good customer for the movie camera attachment lens we were making with the massive-optics process. They were also interested in buying large aspheres for projection television. Germany's largest consumer electronics maker, Grundig, had come to them for optics similar to Advent's that could only be made with our technology. At our suggestion they went to Ellis Betensky for the optical design, leading to a 1978 development program.

After leaving Schneider, Brian and I parted as he took some vacation and I went to a meeting in Vienna. We reconvened in London for dinner before the trip home the next day. Prior to getting together, I called the factory to find out what was happening. The news at that moment was terrible in virtually every area of the business. That evening, I could hardly eat as the awful sinking feeling entrepreneurs sometimes experience set in, causing me to wonder if everything we had built was going up in smoke. We discovered on our return that things were not that bad, and the various problems were satisfactorily corrected. As a result of that call, the Welhams and Howes had an extremely memorable dinner.

In November, Brian had been with us for a little over two years. In recognition of the excellent job he was doing and to involve him more with our outside advisors, he was elected to the board of directors. We were constantly dealing with technology issues in the board meetings, and his direct input contributed significantly to the deliberations.

# 30 Projection Interest Heats Up
and the Delta Lens Development

1978 was the year the major television-makers seriously became interested in projection video. In early January I had been contacted all the U.S. TV-makers in hopes of getting feasibility studies going that would lead to lens sales. We only could talk about the "one-eyed monster" business and the three-tube lenses being made for Advent. The G.E. product was not to be introduced until April 12, so that development had to remain confidential.

The two largest U.S. television makers were RCA and Zenith. Each had over a 20% market share. Magnavox, G.E., Sylvania, Quasar and Admiral all had respectable sales. Although Magnavox was already owned by the European giant, Philips, it operated quite autonomously. Quasar had recently been purchased from Motorola by Matsushita. Sony, the largest of the foreign brands, had attracted much attention because it built a greenfield factory in San Diego to make color TVs. It had the strongest brand name among the foreign makers.

At the January Consumer Electronics Show in Las Vegas, I visited all the exhibits of the American makers to do further research on who the decision makers would be if they were to develop projection products. It was my hope to find and meet them at the show. I had limited success, meeting only David Daly, vice-president of new product planning for RCA, and John Koppier, a new product planner for Magnavox.

Ellis Betensky had been working on a true f/1.3 design for Schneider utilizing four plastic aspheric elements. In January he and Dr. Hohberg, a senior executive at Schneider, and his associate, Ray Muhlschlag, a planning and procurement executive, visited us. Although I had met Muhlschlag on my previous two visits to Bad Kreuznach and Brian met him on his one visit, neither of us had met

Hohberg. Ellis had learned, during his work, that Dr. Hohberg was a very tough fellow with a deep suspicion as to whether we could make what they needed. There were indications it bothered him that the prestigious Schneider Company was forced to come to such a small and unknown outfit as USPL to obtain large plastic aspheres. However, we did receive an order for prototypes, which led to full scale production. As with Advent, we shipped the elements and the assembly was done by them. In Schneider's case they applied a sophisticated anti-reflection coating, utilizing a special vacuum deposition chamber purchased for this particular job. The Schneider lens for Grundig was used in the second three-tube refracting lens projector to come to the market. It was only sold in Europe.

In February, I invited Dave Daly to visit for the purpose of seeing the two Advent systems and a few "one-eyed monsters" that we had set up in a display room. He came with Dr. Jay Brandinger, the vice-president of engineering. Having these two top level RCA executives come to Cincinnati was a good indication of their interest. They were intrigued more by what they might do with a Beta-like lens than Advent's three-tube approach.

Within three weeks, Daly invited us to their Indianapolis consumer electronics headquarters for further discussions. After a technical session resulting in a request for a special single-tube lens a little brighter than the Beta, I was ushered into the purchasing director's office and presented with an agreement to sign, stating they would have an exclusive for any single-tube lens other than the Beta. I told him we couldn't make such a commitment. When pressed further, I said others were already pursuing products with lenses other than the Beta. He had no idea who it was and, of course, we could not reveal the G.E. soon-to-be-announced product. With this surprising development, they decided we should end the meeting and get together again after new thought had been given to how we might proceed.

In early March, Brian and I visited Art Tucker just outside of New York, to see a three-tube system with a single exit lens. This was possible because dichroic mirrors (color filters) were used to position the red, green and blue images for projection through one lens. Tucker had been known for creating innovative projection devices for the military.

We were quite excited about the potential for his approach, although we would later learn there were many technical problems to overcome. He quizzed us as to how he might get large companies interested in it, so I thought about it and wrote a detailed letter suggesting that he license his technology. He never responded. Having decided to do it on his own, he would go on to form two or three companies, each of which would eventually fail. Tucker was an innovative, well-known figure in projection video circles, but a poor businessman. Had he taken my suggestion on how to maximize the opportunity, he probably would have enjoyed far greater success.

On April 25, two weeks after the G.E. introduction, Brian and I visited RCA again. This time we could talk frankly about the pros and cons of single-tube systems vs. the alternative three-tube technology. Also, we now had the first Delta f/1.0 design that doubled Advent's picture brightness with all other conditions being the same. They already had a G.E. projector and were not impressed with its brightness. The conclusion had been reached that the three-tube approach was the way to proceed, but there was no suggestion of a feasibility study or things for us to do as a follow-up. We were disappointed. We thought they would be more eager to pursue the Delta f/1.0 three-tube approach, since it was such a compelling idea.

After much discussion and debate, we made the decision to aggressively promote our projection optics with the American and European TV makers but not the Japanese. The reason for this was that the Japanese dominated the world optical business, and we accepted the popular wisdom that they would never buy lenses from an American company. This was a terrible decision that fortunately we would soon correct.

The Delta I, f/1.0 that Ellis designed—with much consultation and input from Brian—was an incredibly significant development. It delivered nearly twice the light output of Advent and Schneider lenses and required only three elements compared to their four. Ellis' concept was to have the element nearest the projection picture tube be a field flattener that simply prepared the image for one aspheric power element that provided all the magnification. The last element was a thin asphere that, much like a Schmidt corrector, minimized any remaining aberrations. It was brilliant! Years later, the June 16, 1989 issue of *Business Week* was dedicated entirely to innovation and recalled what we had done:

> In 1978, Yoshitomi Nagaoka headed a research team at Matsushita Electric Industrial Co. that was working on a better lens for projection TVs. At the time, such lenses were a clumsy conglomeration of glass elements that required twelve different machine tools to grind and assemble. Then U.S. Precision Lens came out with a novel aspheric design made of plastic. Because of its sophisticated surfaces, just three lens elements did the trick. Nagaoka was shocked: "It was the ultimate in simplicity."

The Delta I represented an enormous advancement in optical performance, but still had flaws. By July an improved version, called the Delta II, was designed. In time, this too would go through a series of upgrades and be followed by many other versions of the Delta concept. Design patents were applied for would later be issued and become of huge value to us. The importance of the Delta design breakthrough and the massive-optic process that made possible its manufacture cannot be overestimated. These creations would define the future of the company.

At the 1978 Summer Consumer Electronics Show in Chicago, we booked a suite at the Ambassador East Hotel. The G.E. lens, Beta II and Delta I were on display with testers that projected resolution patterns onto screens. Also, a couple of "one-eyed monsters" were on hand that compared the bright Beta II to typical inferior f/2.8 and f/3.5 systems. All the major company product-planning and engineering people we knew were invited to the suite to see and discuss what we were doing. Every one of them visited us. While there was interest, they had no sense of urgency and we were given no follow-up things to do. This was frustrating because we desperately wanted to see a major U.S. company make a projector using a Delta lens and thereby establish three-tube systems as the preferred design concept. We were convinced the G.E. single-tube approach would never have enough screen brightness; and even more concerned that if Henry Kloss' new reflective system was accepted as the preferred approach, we would be in trouble.

The big projection thrust at the CES was by the makers of the "one-eyed monsters." A number of the larger ones had displays on the main floor at the McCormick Place convention center, and the smaller ones showed their products in hotel suites. All of them visited our suite. We had two distinctively different groups coming to see us —the methodical product-planners along with engineers from the major consumer electronic companies, and the "wheeler-dealer" single-tube folks that called themselves TV manufacturers. We did our best to keep them separate because the latter group would not have made a good impression on the majors.

During the second half of 1978, nine major companies visited Cincinnati to learn more about what we were doing, what was going on in the market and about our Delta lens development. With sales trips and phone calls, we were fanning their interest throughout this period. Bell & Howell was interested in optics for aircraft projectors. Four Japanese companies—Sony, Pioneer, Mitsubishi and Toshiba — came by. The Sony meeting caused us to rethink our strategy of not pursuing Japanese TV makers. G.E. had three of their top engineering people come to see the Delta lenses, as they also realized the single-tube system did not have enough brightness. Jay Brandinger, Dave

Daly along with Pete Bingham, a bright engineering leader from RCA, visited to see the Delta lenses that by then were shown on a set we had built from a Heath kit. Heath, a seller of electronic product kits for home assembly, was one of our first customers for the Delta II. John Koppier from Magnavox, visited twice, the second time bringing President Ken Mienken and the vice-presidents of marketing, engineering and project management. Zenith had five high- level engineers come to Cincinnati.

Interest in projection television was indeed heating up, but no major company had yet committed to putting a product on the market using the Delta lens.

# 31   A Look at the Consumer Magnifier Market

For some time, we had been searching for major new business opportunities that would diversify our dependence on calculator lenses, auto-focus modules and projection TV lenses. One intriguing area was consumer magnifiers. Bausch & Lomb had a line of such products that were widely sold in office supply and book stores at excellent prices. Over the preceding few years, whenever seeing an interesting magnifier, I bought it. Quite a large collection had been assembled.

We had a problem. Before pursuing this business, a significant amount of product and market research had to be done that we were not particularly qualified to do. Furthermore, it would require a heavy time commitment. As a result of being asked to give occasional lectures to new venture and marketing classes at Miami University, I had become friends with an outstanding marketing professor—Dr. Jack Gifford. After telling him about this opportunity, we discussed the possibility of his designing a class that would study and make recommendations as to whether we should make a USPL line of products. He thought it was a good idea and in the spring of 1978, put together an outline for a project that would be offered as a marketing class for the fall term. We agreed to pay all of the out-of-pocket expenses that would be incurred.

Seventeen advanced marketing students signed up for the class. First they acquainted themselves with USPL and then surveyed sixty companies throughout the U.S. that either made or distributed hand-held magnifiers. Next they looked for potential patent problems. One hundred and seventy-two people in the mid-West were sent an extensive survey that was followed-up with a thorough evaluation of the returns. In brainstorming sessions, they sought new design ideas. They worked with our tooling and production people on manufacturing costs of various designs. Focus interviews were conducted. Packaging, distribution, promotion and warranty questions were studied in depth.

In December we invited the class, along with a few interested professors, to a dinner meeting at the Queen City Club in Cincinnati to report their findings and recommendations to our key people. Also invited were the board and William Liggett, a Miami trustee who was CEO of the First National Bank. The students reviewed all of their findings and presented us with a 151-page report. Their conclusion was that we should not get into the business because it would be difficult to compete profitably with Bausch & Lomb and all the Far Eastern products being imported. There was no compelling reason why our offering would be preferable to what was already on the market.

Having talked often with the students throughout the study, I had already come to the same conclusion. It concerned us somewhat that after all their work they would be disappointed if we didn't proceed, so the recommendation was good news. After dinner, a few of the class members said they were quite worried that their view was not what we wanted to hear, so they were delighted with our reaction.

The students did an excellent job on a "real world" question and learned much in the process. At the same time, with minimal cost, we learned everything we needed to know to avoid a mistake that could have been costly. We were so impressed with their work that eight years later we would collaborate with Professor Gifford on another important research project.

# 32 1978

Early in the year, we were very concerned about a national energy shortage caused by a miners' strike. Companies were asked to make voluntary cutbacks on energy, and we did our best to comply. The strike threatened to have a drastic effect on our business. Dave Hinchman sent letters to President Carter, Senators Glenn and Metzenbaum, and Congressman Gradison, urging action to get miners back to work. He explained that if it wasn't settled soon, our 300 employees' jobs would be negatively affected and business would possibly be lost permanently. Fortunately, before the situation reached that point, there was an agreement. It was, however, a period of high anxiety.

At the close of fiscal 1978 June 30, sales were $6,875,000, up 44%. Net after-tax profit was $788,000, up 432% from our abnormally low fiscal 1977 level, but still 84% higher than the record year 1976. All long-term debt had been eliminated. Semiconductor LED sales were down slightly to $1,448,000. Polaroid business for the SX-70 and Pronto cameras had passed its peak and dropped to $432,000, down 37%. Components and systems sales were down for the first time since the business was purchased in 1970 to $603,000, a reduction of 41%. However, it was more than made up for with $502,000 of new auto-focus module sales to Honeywell. Massive-optic TV projection lens sales jumped 156% to $3,037,000; and now after only two years was our largest business, accounting for nearly half the volume. There was no more talk about "Roger's hula hoop"!

In September, Arthur M. (Stretch) Hoff joined us with the title of Sales Manager. He had been a sales executive with IBM since graduation from Ohio State. Dave had been, for the most part, managing all the market sectors except for projection television, and it had become clear that more sales help was needed. We were not sure then exactly where we would want Stretch to concentrate his efforts, but by mid-1979 he was working on everything but projection television —slowly freeing Dave to help me. In 1986, Stretch became vice-president of sales for optical components. Over the years, he was a major builder of that sector of the business, which became a significant contributor to the company. He retired in 1998.

By calendar year-end 1978, 350 people were employed at USPL. The primary objective for 1979 was to influence a major TV-maker to design the Delta II lens into a projector.

*Henry Kloss, the founder of
Advent and "Father of modern
day projection television."*

*The first consumer projection TV.
Advent model 1000A utilizing
USPL Schmidt corrector lenses.*

*3997 McMann Road in 1976 after
6 additions to the original building.
(bottom)*

— 161 —

Large aspheric lenses for projection television made by the USPL massive optic process.

Pete Cutler joined USPL in 1976 as Manager of Plastic Molding.

Advent Model 750 projection Television introduced in 1976 utilizing three lenses, each containing four aspheric plastic elements from 4 to 5 inches in diameter. Its manufacture was made possible by the invention of USPL's massive optic process.

Mad Man Muntz - the colorful person who said: "Roger, you don't know anything about pricing your products" as he played an infamous X-rated movie on three large screen TV's in his living room (above left).

The 6.1 inch diameter Beta II lens used in most of the "one-eyed monster" projection TV's (above right).

The Business Philosophy of
U.S. Precision Lens.

The
Business Philosophy
of

u.s.precision lens
incorporated

The first rear screen projection television introduced in 1978 by General Electric. Its lens contained three 6.8" aspheric plastic elements.

The first auto-focus camera utilizing the USPL assembled Honeywell optical and electronic module (above left).

Honeywell auto-focus lens module (above right).

Second generation auto-focus camera optic array containing 52 minute lenses shown at 100 times magnification.

Arthur M. "Stretch" Hoff joined USPL in September of 1978 to head the components and systems selling efforts.

The revolutionary Delta II projection TV lens.

# Part Three

*Moving Into International Markets*

# 33 Mr. Itoh Comes to USPL and We Go to Japan

One of the four Japanese company visits to USPL in the latter half of 1978 would cause us to reexamine the decision not to try to sell lenses in that country. On August 2 Mr. Norio Itoh, manager of the image display technology department of Sony Corporation in Tokyo, came to see us with a Mr. Naito from their New York office. Naito was there to assist in travel details and help with interpretation. Mr. Itoh spoke English quite adequately, so translation was not necessary. He was familiar with the lenses we made for Advent and General Electric; he also understood the significant design advantages that were gained by using plastic aspheres. Sony had recently introduced a two-lens, three-tube, one-piece mirror foldout projector, utilizing a dichroic mirror that made it possible to eliminate one lens. The quality was only average, but we were pleased that a company of such stature was in the projection business.

It was clear that Mr. Itoh and his company were enthusiastic about large screen TV, and considerable research and development on the product was taking place. The thing I was most impressed with, as we sat in my office, was his explanation of how Sony liked to work with technology companies on developments that would lead to very long-term business relationships. It was important to him that I fully understand this point. We discussed the Delta development and mutual optimistic views on the future of projection. With no U.S. or European buyers for the new lens yet, I began to reconsider the decision of not attempting to sell to the Japanese companies because they were exhibiting greater interest at that point than anyone else.

I knew from the 1974 trip, that if we were going to try to do business with the Japanese, there would be obstacles quite different from anything we had experienced. Cultural differences, language, business practices, distance and speed of communication were all potential difficulties. In this pre-fax era, we were limited to the use of mail, express mail and telephone. USPL had not yet acquired a telex machine. An important question that was answered quickly was whether to use a Japanese trading company to do the selling. Because of the technical nature of the lenses and the high middleman markups that would have to be added, our conclusion was to sell directly.

Ohio Governor James Rhodes was aggressively pursuing Far Eastern and European companies to choose the state for plant locations. I was on a Small Business Task Force he had assembled for the purpose of finding ways to promote growth. At occasional meetings of the group in Columbus, the International Trade Office he had established in Tokyo was hailed as an important factor in the effort. I contacted James May, Manager of the Export Promotion Program, to obtain information on doing business in Japan and to determine if the Tokyo office could help us.

There were a lot of myths and misinformation regarding conducting business in that part of the world, expounded upon by people who didn't know very much about it. Good advice from individuals with extensive experience was hard to find. I decided we should ignore virtually everything that had been heard, assume (correctly) that we knew nothing, and start educating ourselves on the subject.

Publications on doing business in Japan were obtained from the U.S. Government, State of Ohio, U.S. Chamber of Commerce and other sources. After a thorough study, we assumed it would be a long, slow process requiring much patience. May and a Japanese staff lady who lived in Columbus visited Cincinnati for discussions on how we could work with the Tokyo Office.

We established a selling strategy that had five points. First, pattern our approach as much as possible to the way they did business. Second, provide educational information on projection product and market development. Third, sell the benefits of our aspheric plastic optic technology and its promise for future development. Fourth, demonstrate rapid design, prototype and quality improvement. Fifth, promote ourselves as players in the business for the long term, and develop close relationships at multiple levels.

A fortuitous thing happened as I was leaving the January 1979 Las Vegas Consumer Electronics Show. On my flight was David Lachenbruch, the editor of the highly regarded *Television Digest*, a publication subscribed to by every serious participant in the consumer electronics industry. Lachenbruch was universally respected for his industry knowledge and accurate reporting. I had been acquainted with him for a couple of years, so we visited while waiting for the plane to board. Knowing that we were planning a Japanese thrust, I asked Dave if I could get a tutorial from him on the various TV makers. He agreed, so we arranged to sit together for the nearly three-hour flight. He talked in detail about every company's size, reputation, operating characteristics, technology development capability, distribution system and aggressiveness. I took pages and pages of notes as this recognized authority on the world video business compared all the players. It was an invaluable lesson that gave us a wonderful insight and understanding into the companies we would be seeing.

Promotional folders were prepared containing our capability brochure, the business philosophy of U.S. Precision Lens, a brief history of the company, and a piece on our projection television optic technology. With the aid of the Ohio Trade Office, the latter two items were printed in Japanese.

Philip Jones was the Director of the State of Ohio Tokyo Trade Office, and Tami Hirabayashi was Associate Director. Jones, a Spokane, Washington native, graduated from Harvard, where he learned to write and speak fluent Japanese. He was mentored there by Professor Ezra F. Vogel, a worldwide recognized expert on Japanese culture and business practices. Vogel had written an excellent book on the subject entitled *Japan as Number One*, which sold well in the U.S. and even better in Japan.

Brian Welham and I planned a two-week trip for the last half of March 1979. At that time we thought Stretch Hoff, who had only been with us for a brief time, might become involved with projection video, so he came along.

We requested that the Tokyo Office contact the TV companies' chief engineers and export department managers for the purpose of setting up appointments. Phil and Tami would tell them about USPL, our involvement in projection TV with G.E. and Advent, and about the three of us who would be visiting. Everyone wanted to see us, and appointments were made with the following companies:

Matsushita in Osaka

Sanyo in Osaka

Mitsubishi in Kyoto and Tokyo

Sharp in Tochigi

Hitachi in Tokyo

Nippon Electric (NEC) in Tokyo and Osaka

Toshiba in Tokyo

Sony in Tokyo

General Corporation in Tokyo

JVC in Ibaraki.

On March 15, we departed for Japan with many promotional folders and Delta II samples. At each company, we were greeted by five to fifteen people for a meeting that usually lasted about three hours. The presentation was always the same. In Japanese, Jones briefly described his role for the Ohio Trade Office and introduce them to USPL. I then told a little about our history and commitment to projection lens development. Jones interpreted if necessary; but when when I spoke slowly, without the use of colloquialisms, it was not necessary. I emphasized the uniqueness of the massive-optic technology, our view that projection would become a big business, and our goal of seeking long-term relationships that would be beneficial to both our companies.

Usually there would be a response from the leader about their company, its interest in projection, and what they knew about USPL.

Typically this was be followed by a discussion on what was publically known about various companies' activities, market trends and forecasts. With this out of the way, the technical discussion began. I gave a brief description of the Delta lens and its attributes. When the questions came Brian take over: not only providing the answers but also giving a tutorial on how to use the lenses, how they related to the rest of the system, and the various configurations that could be designed. We were doing far more than just selling lenses—we were showing them how to make projection television systems. In every case, the Japanese were intensely interested in what was being introduced to them. After departuring, Phil Jones reported side comments overheard and his observations of the meeting, which were nearly always positive.

We had been warned of a very big cultural difference between Tokyo area companies and those in the Osaka, Japan's second largest city. In Osaka, they were far more conservative and tended to buy components from companies nearby. Matsushita was the largest consumer electronics maker in the world, with most of its TV R&D and manufacturing in or near Osaka. It sold its products under the Panasonic, Technics and National brand names. Our sense was they viewed us skeptically. Sanyo had the reputation of making average goods and depended on finding new ideas in other makers' latest products. Sharp was considered aggressive, cost-effective and innovative. All competitors watched Sharp closely. Toshiba was a huge company with interests in many areas, although in consumer electronics it was slightly smaller than Matsushita. The same could be said for Hitachi. Mitsubishi was considered the leading projection maker in Japan. They used a cumbersome Schmidt system much like Henry Kloss's first offering. We sensed they were somewhat wary of us. NEC showed considerable interest. JVC was a surprise when we reached their plant in Ibaraki; in front of the building was a large ceramic dog and victrola that was the famous early RCA logo. We then learned that before World War II it had been a joint venture with RCA called Japan Victor Corporation. They were attentive, but not yet doing much in television. General Corporation, a small local maker, was obviously not a good prospect.

We had been sending Delta II samples to Toshiba since their October visit to Cincinnati and there had been a considerable amount of communication. When visiting them we were delighted to learn they planned to produce a 45-inch, one-piece mirror foldout projector for the Japanese market using Delta II lenses. It was also to be sold in other Pacific Rim countries, but not the United States. Toshiba was the first major company to purchase Delta lenses. Unfortunately, because the brand was not strong in our market and the product would not be seen here, there was low expectation that the U.S. companies would be competitively influenced.

When the procurement manager was asked how many they planned to buy, he emphatically said, "lots of lenses!" An initial order for over 6,000 lenses with a value of approximately $300,000, was issued. Along with the order came pressure to have it expedited because they wanted to be the first company serving the Japanese market to have a three-tube refracting optic projector.

An afternoon visit to Sony Corporation was extremely exciting for us. While Sony held the fourth or fifth largest TV share in Japan, it had by far the biggest foreign business. It was a company with an unusual spirit that saw itself as a world player; as such, was quite different from the others in the way it did things. They had a recognizably more open way about them, probably as a result of Co-Chairman Akio Morita's move to New York in the early 1950s to learn how business was done in the United States and to establish sales offices. A new television plant had been built in San Diego that brought the company much notoriety and favorable press.

We met at the Sony headquarters building in the Shinagawa section of Tokyo. Mr. Itoh assembled an impressive group of high level development and manufacturing people. One was revered senior scientist Akio Ohkoshi, a shy man whose specialty was electronic circuitry. He had joined Sony when it was a tiny company, becoming employee #200, and gained fame in 1967 as one of the two principal inventors of the Trinitron picture tube—a feat that earned Sony an Emmy Award and, more importantly, the reputation for having the industry's best picture quality. Another participant, Takizoh Shioda, was the man responsible for developing and engineering new projection products. Tatsunori Toki worked for Shioda and coordinated all details of procurement and scheduling. Takuji Inoue, an assistant to Ohkoshi, and Hiroshi Horiuchi, an optics specialist, were also there. They were very interested in the Delta II, all the insight Brian could give them on performance, and our views on the potential for improvements.

As this exhilarating meeting came to an end and we were about to depart, Toki said that if we were free that evening Mr. Ohkoshi would like to entertain us for dinner at his club in the Roppongi area. The three of us, Jones, and at least five of them had an enjoyable time getting to know each other that evening. It was clear Mr. Ohkoshi, who soon would become Dr. Ohkoshi, took a special interest in us. Later Jones concluded that we had a very good chance of doing business with Sony.

When the trip ended, we were very pleased to have received an order from Toshiba so quickly and believed that Sony, as well as Hitachi, were excellent prospects. Everyone else, except General and JVC, were good prospects. All were evaluating the Delta II with its sensational light output. Unfortunately, it still had excessive flare, causing a slight halo look around dark objects and reduced contrast. Brian and Ellis were working to correct the problem.

Mr. Shioda visited USPL on March 30, our first day back. He obviously had to be in the U.S. for other reasons. We were encouraged by his changing plans to take a firsthand look at us so quickly.

Phil Jones contacted me within a week to say that Mr. Masaru Ibuka, the Senior Co-Founder of Sony, might like to visit us.

I responded that we would be delighted and honored if he would come. Mr. Ibuka, although less flamboyant and well known than co-founder Akio Morita, was recognized for superb managerial and technical ability. Later we would learn he was the company's strongest projection TV advocate.

Unfortunately, there was one potential problem with his visit that I felt needed to be explained up front. The details of the large aspheric lens massive-optic manufacturing process were never revealed to anyone because our strategy was to patent the unique lens designs it made possible and keep the process itself a secret. I asked Jones to explain carefully to Sony that our internal secrecy agreements and policy prevented a complete tour. Everything else would be shown to him, and we very much looked forward to his visit to Cincinnati. Mr. Ibuka never came. Perhaps he couldn't work it into his schedule; but I always assumed that when learning he would not see what probably was of the most interest, plans for the visit were abandoned.

Shortly after returning from Japan, we were visited by Yoshihiko Metsugi, of NEC. Kloss Video, Henry's new company, was now in the process of gearing up for production utilizing his improved Schmidt-type reflective tube. We made the corrector plates. NEC acquired a license for the technology, and Metsugi came to arrange the purchase of correctors that would be produced from the Kloss tool. Although this was good molding business, it also represented competition for our Delta refracting optics.

In June, Mr. Murakami and Mr. Marima of Hitachi came to Cincinnati, following an invitation we had extended in Tokyo. Their television people had a top reputation for circuitry design that was widely admired by others in the industry. In fact, Hitachi and G.E. had announced their resources would be pooled to form a new company called Television Corporation of America. Not long thereafter the U.S. Justice Department denied approval for the venture, causing each to go their separate ways. Now, Hitachi was looking for ways to move meaningfully into the U.S. market, and projection video was thought to be an avenue.

The Toshiba Delta II assembly details were being worked out, and a production line was set up in the converted second molding room. There was hesitation to invest heavily in it because we were building a new TV lens factory a couple of miles away that was due to be finished in the next year. The Toshiba order was being rushed for shipment in July.

Sony had asked for flare improvement in the Delta II, causing Ellis and Brian to make surface modifications and design a fresh version called the Delta II-M that required a new housing. The housing molds were very expensive, so we only made prototypes for initial evaluation. The cost-effective solution to the problem called for tuning the original version.

On September 29 Brian and I departed for Japan with samples of the improved Delta II and a prototype of the Delta II-M. Sony was the first stop. They agreed the tuned samples were better, but there was a lack of enthusiasm about the changes. We did not show them the Delta II-M, hoping to avoid the costly retooling.

Toshiba was the next visit. We had already heard there were some issues with a few of the production lenses that had been shipped. Phil Jones had also received a call saying it was important to get there quickly for discussions. On September 2 we arrived at the Fukaya factory for what would be one of the toughest and most embarrassing meetings Brian and I ever attended.

We, along with Jones, were ushered into a large conference room that had a very long table. There were at least fifteen Toshiba people on one side with the Projection TV Project Manager in the middle, which denoted that fact that he held the highest position. The others were seated to his left and right in descending rank. I was seated across from him, with Brian on one side and Phil on the other. The crossed American and Japanese flags were on the table between

me and their leader. As tea was being served, gracious introductory remarks were exchanged that did not yet deal with why they so urgently
wanted to see us. What followed was simply awful.

The project leader said that at this particular plant, they had
over 2,000 suppliers and that two of them were not doing a good job.
USPL was one of the two, and, frankly, far worse than the other. There
were serious quality problems with our shipment. He went on to say
that since they knew us and had confidence we were good people, the
only conclusion that could be made was the problem was caused by
inferior American workers. Then he detailed the lens defects that had
been found. Some of the housings did not fit together properly, some
of the lenses were loose in the housings, screws and other pieces of
hardware were missing in many instances, and a few screw threads
had been stripped. The leader was remarkably calm as he delivered the
overview; then his associates, in excruciating, laborious detail, provided the statistics from their incoming inspection. The frustration of
the lowest level fellow at the end of the table was much easier to read.
According to Jones, he was furiously demanding in Japanese that the
three of us get out in the warehouse immediately to fix the problems
he had been looking at for days. Fortunately, he was signaled by his
superiors to quiet down. The meeting took about three hours, but it
seemed much longer.

If it had been an earlier time in history and I had come from
their culture, the proper thing to do at this point would have been to
fall on my sword. Unfortunately, we had a serious problem that
demanded a satisfactory action plan. It was now up to me to respond
to their presentation. First, I humbly apologized at length for all the
trouble we had caused and told them we were deeply embarrassed. I
made a further number of points: this was not typical USPL performance; we would correct all of the problems; this would never happen
again; and, finally, the problem was not a result of inferior American
workers but rather a failure of our top management to provide the

right training to the people who assembled the lenses. For this, I personally took complete responsibility.

The Toshiba leader and his top lieutenants amazingly thanked us for the response, adding they looked forward to receiving lenses of flawless quality from us in the future.

We were humiliated. It was the worst meeting I ever remembered attending. Brian and I could hardly believe we had failed so miserably, and we reiterated to each other that it must never happen again.

After visiting Hitachi, we moved to Kyoto to call on companies in the Osaka area. The plan was to return home from there. While seeing Matsushita, NEC, Sanyo and Mitsubishi, I continued to think about and be nagged by Sony's indifferent reaction to the tuned Delta II. I finally told Brian we had to extend the trip by a day, go back to Tokyo, and show them the Delta II-M prototype. He agreed. We called Mr. Horiuchi to arrange a late afternoon meeting following our arrival on the bullet train. Horiuchi put the II-M on a tester that projected a resolution target and compared it to the II on an identical tester. The new lens reduced flare and provided a crisper image with more contrast; demonstrating a dramatic quality improvement. He excitedly called Mr. Shioda and others, asking them to come see what we had brought. Everyone was impressed.

This last minute return visit would lead to our receiving very large orders that were used in 50-inch and 72-inch two-piece coffee table projectors introduced in mid-1980. Although it meant an expensive retooling, we were delighted at last to have a major TV manufacturer using Delta lenses for sets that would be sold in the United States. We knew that when the Sony projectors were put into the market, other makers would be pressured to have a competitive offering.

The first thing we did upon returning was stop massive-optic production and assemble all the associates for a meeting about the disastrous Toshiba visit. All the defects were reviewed in detail and they were informed about the customer's assumption that the problem was because the inferior American worker. Everyone was in shock. I concluded by saying that I told the Japanese manager it was not our American workers' failure, but rather mine for not providing adequate training and that it would never happen again. The people in that meeting were as embarrassed as we had been. Training changes, assembly procedures and inspection improvements were immediately instituted as we all set out to redeem ourselves and prove the American worker was not the problem. From that point on, Toshiba received lenses with flawless quality.

Within three weeks, Paul Stancik, head of U.S. procurement for Sony visited Cincinnati along with engineering and quality assurance people to assess our capability. Two weeks later Shioda, Horiuchi, a couple of engineers, and Stancik's assistant came to establish final specifications for the Delta II-M. It had been fourteen months since Mr. Itoh visited USPL, causing us to change our minds about pursuing business in Japan; now we were quickly establishing a very significant relationship with one of the great companies in the world.

# 34 The Long Walk in East Berlin

In late August 1979, Brian, Dave and I traveled to Europe for visits to Schneider in Germany; Philips in the Netherlands; and Barco in Belgium. This trip coincided with the huge West Berlin European Consumer Electronics Show called Funkaustellung, so we decided to attend. There was a problem, however, because this last-minute decision was made after all hotel rooms in West Berlin had been booked. The travel people suggested we stay at a fairly new hotel, the Metropole, in Communist East Berlin near the infamous wall. Brian balked, saying he wouldn't go because the East Germans could well know of his military optics work at Bausch & Lomb and have a special interest in him. Dave and I went ahead while he returned to the U.S.

We landed shortly before noon and proceeded to the Metropole. At that time there were two ways to pass through the wall. One was by foot at Check Point Charlie, and the other was by rail to the Friedrichstrasse Station. We chose the rail because it was close to the Metropole.

It is difficult to explain the emotions one experienced when passing through the Berlin wall. It had a foreboding appearance, with guard stations holding visibly manned machine guns spaced for complete firing coverage. When the train reached the station, we were met by armed Communist soldiers who directed everyone to a line where passports were inspected, papers were issued, and a certain amount of East German money had to be purchased. From there it was a short walk to the hotel.

The Metropole was built for the express purpose of attracting Western currency. East Germans were not allowed to enter unless invited by one of the hotel guests. The local money that had been purchased was not accepted and could not be reconverted when exiting. There was no doubt that listening devices existed in all the rooms.

Food, beverages and accommodation quality were at Western standards —quite different than would have been found anywhere else in East Germany.

Dave and I had a dinner appointment with a Zenith executive in West Berlin on our first evening. We decided to go back through the wall at Check Point Charlie, so a taxi was needed. The lady clerk at the Metropole desk informed us that a four-hour lead time was required and placed the order. I asked her the location of a certain restaurant in West Berlin. She didn't know the answer because she had never been there—another reminder of what a different place this was.

The taxi arrived at the appointed hour just as darkness began setting in, and drove us the twenty-five or thirty blocks to Check Point Charlie. One guard armed with an impressive automatic weapon looked at our entrance papers and said, "No—come in Friedrichstrasse, go out Friedrichstrasse." Dave started to argue with him (I don't argue with people in uniforms who have guns), and I rushed to retrieve the taxi. It was too late. We watched its tail lights disappear into the early darkness.

A light drizzle was in the air as we started a long, memorable walk to the Friedrichstrasse Station. There were no street lights and no cars on this drab boulevard that once had been the center of Berlin's bustling city life. Occasionally we saw a light bulb dangling in a building entrance, dimly illuminating two or three shadowy figures watching us pass. Never had we been in such a defenseless, helpless position, and we could well imagine that if anything happened no one would know how to find us. To say we were fearful would be an under statement. Nothing happened, and with great relief, we finally made it to our dinner meeting.

As a result of that wall and the two different systems of government, the contrast between East Berlin and West Berlin, was far greater than anything I had envisioned. At the time, I wished my family and everyone at USPL could have experienced this visit because there was never a clearer demonstration of how lucky we were to live in a free society, and particularly, in America.

After two nights at the Metropole, Dave and I took the East German currency we had accumulated to a local store to buy something because there was nothing else to do with it. The place we went into had virtually nothing on the shelves. All that could be found was a cheap perfume that, frankly, didn't have a very good aroma. Like everything else we had seen in East Berlin, it was another example of the benefits Communism had brought to their people.

# 35 Competitive Advantage

There has been one frequently recurring question asked about USPL over the years: how did we become such a successful and dominant projection lens supplier to the Japanese electronic companies? It is a good question, in light of the fact there is such a strong optics industry in Japan and that few American companies, particularly in the late 1970s and the 1980s, had been successful selling American-made products there. The answer is not simple because it has a number of components.

Already discussed was the important mindset that we would not pursue these companies with preconceived notions, but rather learn as much as possible about how they did business and adapt our approach accordingly. Clearly, we had unique design and manufacturing technology that was significantly different from anything available elsewhere. The massive-optics process made possible the use of large plastic aspheres in imaginative ways that had not been conceived of before. It allowed for such a major departure from conventional lens design that electrical engineers, with limited optics experience, had difficulty grasping its importance.

The early USPL projection TV lenses were brighter and less costly than glass alternatives, but had limitations. Image sharpness and contrast were areas Brian and Ellis aggressively sought to improve. In doing so the focal lengths were shortened, allowing makers to fit all the components into smaller cabinets that were less visually intrusive to the home. We developed procedures to build massive-optic tooling quickly, so prototypes and samples could be produced with comparatively short lead times—a capability greatly valued by our customers and a problem for our competitors.

In the mid-1970s in Los Angeles, Vivitar Corporation hired two lens designers to augment the work Ellis Betensky was doing. It was part of a grand plan to make the company an integrated camera and lens maker. When they finally determined the program was too ambitious it was cut back, eliminating the need for Mel Kreitzer and Jacob Moskovich, the designers. At the same time in 1979 our need for optical design was exploding, and TV makers were looking for improved performance and configuration advantages. Over the years, Ellis had his offices in Stamford, Connecticut; Toronto, Canada; Tel Aviv, Israel and New York. While we were delighted with his work, it was becoming more difficult to get the volume needed on a timely basis. He recognized this, and together we concluded that the Vivitar switch in strategy created an opportunity to bring Mel and Jake into his operation. We also decided it would be ideal to have them in Cincinnati, where they would work as part of independent Opcon on USPL's premises.

Mel Kreitzer, a native of South Africa, had earned his doctorate degree in optics at the University of Arizona in 1976. It was one of the two U.S. schools most renowned for optical programs. Jacob Moskovich, a native of Russia, graduated with a master's degree in physics and optical engineering from Chernovsty State University in 1973, and migrated to the United States in 1974. Upon their departure from Vivitar in August, they set up an office in a separate area of our headquarters building. With a Univac computer, the technically advanced Opcon optical design program, and these experts working with Ellis, we had tremendous capacity to do optical design. It was a fundamental strategic move that, much like acquiring the mold-making company in 1972, gave us powerful internal capability to bring against design challenges on short notice. When a customer asked for a special design or Brian came up with a new idea, it was just a short walk to the Opcon office to set it in motion.

With quick tooling and rapid design capability, we were able to aggressively ratchet the TV lens technology, making our existing products obsolete quickly. This frequently left competitors working to respond to out-of-date creations. In addition, patents on the new designs were applied for whenever an important feature was found. Our strategy of keeping the manufacturing technique secret and patenting the designs worked beautifully.

There was yet another aspect of our effort that differentiated us from the Japanese lensmakers. Brian, and others of us to a lesser extent, understood how the electronics of television systems interacted with projection optics. This gave us the ability to teach customers how to design the devices to interface with what we provided. In the early years of projection video, our sales calls often appeared to resemble seminars on all the things one needed to know to build a set. Customers everywhere valued USPL's ability to provide this kind of education. Of course we were learning a great deal from them in the process, and that increased our knowledge bank.

In 1979, a Magnavox executive asked us to present a paper to the Institute for Electrical and Electronic Engineers or IEEE, one of the largest and most prestigious technical societies to which virtually all important television engineering people in the world belonged. Brian and I saw this as an opportunity to write a primer detailing the information we had been imparting on systems and projection television optics. We thought this would free us from having to continually go over the basic information again and again as new people came on the scene.

On a long flight to Japan in September of 1979, we jointly organized and wrote the paper. The more important technical detail was provided by Brian. That occasion provided an indelible memory when the pilot of the Boing 747, two hours past Alaska, informed us the headwinds were so strong we had to go back to Anchorage for refueling. This was not an uncommon occurrence on this route during the fall season, especially with the planes used in those days. A number of other flights had the same problem, so at least a half a dozen jumbo jets were on the ground for their fill-ups. People were allowed to get off the plane for some fresh air on the Anchorage tarmac, and a few lit cigarettes—not a compatible act with aircraft refueling. A frantic pilot rushed to have them extinguished and had us immediately reboard. It was hard to imagine anyone could lack enough intelligence to smoke under those conditions.

On November 13, I delivered Brian's and my ten-page paper, entitled "Developments in Plastic Optics for Projection Television Systems," to the IEEE at its Fall Consumer Electronics Conference in Chicago. It covered systems considerations, signal sources, CRT design, x-rays, heat factors, phosphors, internal reflections, screens, and of course, all the important things to be taken into consideration when choosing plastic optics. The effort was well received and did exactly what we had hoped in providing basic projection education. Over 1,000 copies were distributed in the U.S., Japan and Europe, and is further enhancing USPL's importance to the development of this new technology.

The combination of our educational efforts, plastic asphere capability, the ability to quickly make design modifications, and short tooling time added up to an important competitive advantage. Japanese glass lensmakers never made it a point to understand the total system, and their designers were often university professors not closely tied in to the manufacturers, creating a situation we thought had limitations. They were at a disadvantage with cost and prototyping lead time, being limited to glass spherical elements. We had already developed great respect for Japanese companies' ability to adjust to adverse conditions, and therefore assumed that if they were not tough competition at the moment, they would be in due time.

Although the importance of the yen/dollar relationship had been understood when it came to travel expenditures, the full realization was just becoming clear that an undervalued yen and overvalued dollar had immense consequences for our ability to be competitive with the Japanese. A greatly undervalued yen had the potential to make their expensive five or six-element, glass lenses cost competitive to our plastic offerings that were being sold in dollars. When it reached a low of 251 to the dollar in 1979, virtually all American business people trying to sell there or compete against Japanese imports thought it was extremely undervalued. The yen/dollar relationship was not a competitive advantage but rather a huge disadvantage for us.

Our visits to Japan became frequent and followed the same pattern. Calls would be made on customers at multiple levels. With the top managers, meetings were of a more general nature involving talk about market trends, new developments at our respective companies, and an overview of what we were working on together. With the project engineers, the sessions were very specific. On every visit we made it a point to have improved versions of designs, better prototypes or new design concepts. Almost always, the key engineering players and very high level managers were invited to dinner in the evening. In a departure from what we had become accustomed to in the U.S., more often than not they insisted on entertaining us. On our return, follow-up letters thanking them for their hospitality and answering open questions were sent to nearly all the people seen.

The answer to the question of how we became so dominant was, in spite of the yen/dollar relationship, all of the above. Superior products and continual improvement were not enough.

Strong relationships at multiple levels and mutual trust also were a critical part of the equation.

# 36 1979

As this was written in 1999, we looked back on seventeen years of comparatively low inflation. Only in 1990 during this period did the consumer price index (CPI) exceed 5%. In that year it was only 5.4%. There have been no severe recessions since the early 1980s. The average bank prime interest lending rate has been in single digits for the last fifteen years with the exception of 1989 and 1990, when it was slightly over 10%. From 1985 to the present, the United States economy has, by historical standards, been remarkably strong and stable.

It has not always been that way. Business cycles taking us from prosperity to recession and back to prosperity were accepted as normal business phenomenon. From the early 1970s, we found ourselves having to adjust with regularity to economic changes that brought employment buildups and cutbacks reflecting what was being experienced by our customers. Calculators and projection television were new consumer products, so there were distribution pipeline filling variables as well as forecasting misjudgements that combined with general economic swings to create instability.

In 1979, the business environment became particularly turbulent. The inflation rate jumped to 11.3%, and the prime interest rate reached a high of 15.6%. We had never seen anything like it. There was deep concern about what President Carter called our "economic malaise." We were concerned about what all this meant to us at USPL. On February 12, I sent the following cautionary memo to our management associates indicating volatility concern and the reasons why a strong balance sheet was important. It conveys the mood at the time.

---

**INTEROFFICE MEMO**

TO:  Management Group    DATE:  February 12, 1979

SUBJECT:  *Short-Term Strategy for USPL*

Frequently, you have heard us describe our business as being *volatile*. This is because the bulk of our product ends up in consumer products—particularly new technology consumer products such as photographic equipment, T.V., contemporary toys and semiconductor products such as calculators. And because we are involved in a new technology that competes with other new technologies, volatility of our activities must be understood and, to the extent possible, used to our advantage. At the very least, we don't want it working against us.

Therefore, if we accept as fact that this is a high technology, volatile and consumer-oriented business, we must develop *strategies* that address these factors and *current economic conditions*. I believe that no one in our company is smart enough to predict very far in advance, and in the absence of hard contracts, how these factors will affect us. Therefore, we must try to *position ourselves to handle downside moves* and, at the same time, *take advantage of upside opportunities*.

Our first conclusion is that *we must be strong financially* for two reasons. In the event of downside volatility, either with customers or the economy, we must have strength to weather storms without having to make moves that impair our ability to achieve long-term objectives. Secondly, in the event of substantial upside opportunity, we need the financial strength (strong balance sheet) to underwrite fast growth without greatly weakening that strength.

Our longer-term strategy is clear. It is touched upon in the philosophy. We want to be the best outside supplier of plastic optics, and we want to lead with the technology. However, we only want to do these things where it is clear there is a need and that we can provide it at a reasonable profit from the base of a strong company.

Long-term strategy can only be achieved with the accumulation of a series of essentially successful short-term strategies. Our task is to develop and execute the short-term strategies in a way that is not inconsistent with the longer one.

In November, we met at Kings Island and reviewed scenarios for the next two years. Out of that came capital spending options with all signals being that we move forcefully along optimistic lines. Now what is different, and what does it mean to us?

1.  Government moves to turn the economy to the right track have thus far shown no positive effect.

2.  The oil and energy situation is more confused and more bleak—particularly in light of the Iranian situation.

3.  Consumer debt is at its highest level ever, which suggests it can't keep climbing, which further suggests a slowdown in consumer spending.

4.  About a month ago, we saw hesitation and slowing in Massive-optics, and now we are seeing the first signs in the basic business. At the moment neither are really weak; but for the months ahead, we do not see the pressure on orders that we have been used to.

This all means *caution* with no change in strategy. We are going to have to run a tighter ship and better justify every step we make in light of our commercial opportunity because there may not be so much of it for awhile. All of this can change very quickly in a volatile business—either way—but this is how we see it at the moment.

Our situation is not bleak, but at the moment it is not as strong as it has been. We will undoubtedly have a poorer second half, which does not please me at all. The good news is our basic financial position is strong, and our longer-range opportunities look as good if not better than before.

Following the March Japan trip, for the remainder of the year we had twenty-eight visits by twelve major television makers to USPL. While a few began to execute product plans, the others were trying to decide how to proceed. Having them visit our facilities was advantageous in a variety of ways. It gave us a chance to exhibit capability, demonstrate Delta products as well as prototype lenses on resolution testers that showed the progress being made, and we were able to have open ended, highly-focused discussions on all aspects of projection television. Usually everyone came to know each other better at dinners the evening before or the evening of the meeting.

When fiscal year 1979 ended June 30, sales were $8,818,000, up 28%, and net after-tax profit for the first time passed the one million dollar mark at $1,011,000. There was no long-term debt. The LED semiconductor business had rebounded from being down in 1978 to a record $1,807,000, up 25%. It was our last important year with Polaroid, with sales of $251,000, down 58%. Components and systems were up 41% to $852,000. Auto-focus camera modules were up 180% to $1,404,000, making it our third largest business segment. Massive-optic television business was up 9.3%, to $3,320,000, comprised of Advent, G.E. and Beta lenses. It was not until the second half of the calendar year 1979, which was fiscal year 1980, that Delta IIs were shipped to Toshiba.

We did not have a human relations department, but rather one person who did the necessary paperwork when associates would come or leave. That person interfaced with accounting on financial matters and the various departments that did their own hiring. Many of the people that came to USPL were referred and recommended by existing associates. With over 350 associates, it was clear we were overdue for professional help in this area.

One evening I was telling my friend and neighbor, Ward Withrow, about the problem. The H.R. function ultimately reported to him at the Fifth Third Bank, so he was able to provide good advice on the matter. He urged me to contact Jerome J. Behne, who had worked for him for a number of years before moving to a bank in Florida. He had recently returned to Cincinnati and Ward, who held him in great regard, thought he would be a perfect fit for our needs.

Jerry Behne graduated from Thomas More College with a degree in history. At Xavier he earned a master's degree in education, and a master's degree from the University of Cincinnati in labor relations. When I met him I was very impressed and thought he had human relations insights similar to what we had seen in Charlie Arnold, whom Ward had recommended as someone that might help us years earlier. Dave was also impressed. It was our good fortune to convince Jerry he should join USPL. He started December 18, 1979, with the title of personnel manager.

Soon after arriving, Jerry was doing all the things good H.R. people do, such as tidying up our employment files, improving policies and providing services to associates on the myriad of personnel issues that came along. In addition, he set out to methodically increase the skills of our workforce through better hiring, internal training and external education programs. He was able to do it with sensitivity and gained trust of associates at all levels. It was not long after he joined us that everyone wondered how we were ever able to operate without him. In 1986 he became vice-president, human resources. Throughout his career, Jerry was an outstanding contributor to the building of USPL. He retired in 1998.

As the year 1979 came to an end, the bad news was the economy was terrible. The good news was that in spite of the bad news we had Sony, a world renowned new customer, buying the Delta II-M, and it had become obvious that other major TV makers were planning to make projection television receivers utilizing lenses from U.S. Precision Lens.

# 37

## People: Joys and Disappointments

It would be misleading to infer that building USPL was easy. All successful businesses start with a good idea that required capital, facilities, sales and continual improvement of the idea. It is all propelled by people; and, although perhaps obvious, almost everyone who has built a good business will tell you that finding, developing and managing an organization is the toughest part of it all.

When we moved to McMann Road, we were able to hire excellent people. They learned their jobs quickly, and many were able to take on greater responsibilities. We promoted a number into lead positions far more quickly than we thought possible. With a few exceptions of course, virtually everyone who joined us took pride in their work, cared about doing a good job, and considered USPL "their company." It was wonderful to watch associates move up from within. We had many (and today have many) who have grown to make enormous contributions to the company's good fortune. This was the joy.

The sadness was to sometimes see the very individuals who moved us forward stop growing and begin to hold us back. It usually seemed unfair to them as we moved others ahead. I always worried that occasionally the problem might be some fault of ours in training or support—and, in a few cases, it probably was.

We had a fast-growing business with a technology that required continual product performance improvement. It was always a problem to find enough moldmakers, train enough molding technicians, and find the right people within or on the outside with skills that were needed. Add to this the fact that we were frequently changing the way we were organized in quest of a more efficient and smoother operation. Slow-growing companies have the same problems. However, such problems are much more intense in fast-growing ones. In our case, with new associates at all levels, some hiring and promotion mistakes were made that would have to be corrected.

Over the years, we would have many highly talented people who were critical to our success. They usually had a few qualities that were less than we wished for. Anyone who was ever a part of USPL management often heard me refer to them (or defend them) as our three-sided people. It was meant to be complimentary. In a world where we would have hoped for people with four desirable sides, ours frequently had only three. The top-flight large corporations could usually get the well-rounded four-sided ones; but given our size and maturity in the early days, with a number of exceptions, we could not. I always accepted this and actually took pride in the fact that these people could succeed in our company. It only required that the rest of us with, of course, our own limitations understand, make accommodations and work around the one weak side. The talent that blossomed as a result of their good sides, was often amazing.

Like all successful companies, USPL was always in a constant state of rapid change. In memos to our people, and in associate meetings, I frequently talked about why it had to be this way, and why it was part of a natural progression. I would emphasized that as individuals we had to change and improve too, or the enterprise would outgrow us. Throughout the years we heard people talk about the "good old days." In some ways there were things about earlier times we missed. But, in general, the "good old days" were never as good as people like to think because we tend to have selective memories. When we saw a person actually become unhappy because things were not like they were in the "good old days," it was a pretty clear sign the company had outgrown them.

Fortunately, we were blessed with many people that continued to grow, make major contributions, and find success for themselves. Those that didn't make it were far fewer in number. Along with the overall success of the business, nothing gave me greater satisfaction than to see its builders do well too.

# 38 1980: USPL Is 50

This year marked the fiftieth anniversary of Herman Buhlmann's founding of the enterprise, and ten years since I purchased it from Bob Mayer. From 1970 to 1980, it had multiplied in size fifty-four times, and the future looked extremely promising, driven principally by the opportunity to grow the PTV or projection television market.

Sylvania, Zenith, General Electric and Hitachi each had development programs for new projectors utilizing a new shorter focal length lens called the Delta IV making it possible for them to somewhat reduce cabinet dimensions. The large size of PTV models was a negative factor in the minds of too many buyers, so our design work then and in the years that would follow included considerable effort to shorten the light path, allowing external measurements to be shrunk. Picture quality and cabinet size were the two product features that most challenged us and all the makers.

The Delta IV was a 114mm lens sold as an f/1.0. In its final form it was actually an f/.94, causing a favorable 13% increase in brightness over what would have been expected. We were amused to learn of a Japanese competitor's bewilderment when, while prototyping a many-element glass lens to the same advertised specifications, he found the Delta IV to be significantly brighter. It was not our intent to deceive anyone in failing to promote the additional benefit—we simply overlooked the advantage.

Projection video was beginning to divide into two distinct market categories. The one receiving the most attention was the consumer, or the home entertainment market. The other was classified as institutional, indicating devices used for a variety of business and specialized entertainment applications. These were projectors that were hung from a ceiling or placed on a stand projecting an image to a separate screen. Except for the aircraft versions, the screen sizes were from four to ten times larger than consumer models. They were used in boardrooms, training centers, conference facilities, auditoriums, casinos, airplanes, simulators and upscale home theaters. Specialized versions also were built to visualize a large number of complex variables in technical control centers such as those used by the space program, railroads and power plants. Frequently these involved multiple projectors and screens. These units produced images that often included data requiring higher picture resolution and more even light distribution on the screen. Initially, the companies building institutional projectors were Sony in Japan and Barco in Belgium. The lenses were more complex and costly because a greater number of elements were required, manufacturing tolerances were higher and production runs were shorter. In the years that followed, over a dozen more companies would produce for this market segment, and we would design scores of different lenses to serve the market. It became an important part of our business, and in some years accounted for nearly a third of USPL's dollar volume.

The continuing problem of having enough space and production capacity was becoming particularly acute. We foresaw this in 1979. After a study of how to expand the existing factory indicated undesirable options, five acres of land were purchased a couple of miles away on Bach-Buxton Road and Miller-Valentine Company of Dayton was engaged to design and build a 33,000 square-foot plant that would produce television lenses exclusively. A secondary reason for having a separate location was that in the event the television business was lost to new technology or competition, we would have a discreet asset that could be sold.

In May, all the massive-optics lensmaking and assembly equipment was moved to the new factory. With a vastly better layout, adequate space and a production management change, quality as well as product flow improved dramatically.

There had been three different production managers in charge of the massive-optics division, and at about the time of the move, it became clear that yet another was needed if progress was to be made. It was not an easy job because, with a new and evolving technology, product changes were frequent and processing modifications were continually being made with the discovery of new techniques. Dave Hinchman was becoming increasingly involved with the projection television business, particularly the production aspects, and believed strongly we needed a leadership change. On May 13 he installed Leonard Kosharek as the operations manager of massive-optics manufacturing.

Len had joined us in March as a manufacturing engineer in that area. He was a native of Milwaukee who had taken engineering and production management courses at the University of Wisconsin Technical College and the University of Cincinnati. With work experiences as a process engineering technician, plant engineer and manager of manufacturing engineering for Johnson Controls, he had a particularly good background for the demands of his new job. When Len assumed the position, improvements came swiftly. His production and technical skills, combined with a flexible attitude, gave him the ability to contribute significantly to the building of our television business until 1987, when he departed to buy a consumer electronic retailer.

In April of 1980 I was at a meeting in Madrid, Spain, where two unsettling things happened. Iran had been holding American Embassy diplomats hostage for months, to the frustration of the American people and our friends around the world. President Carter sent a contingent of Marines in to rescue them. Tragically, the attempt was a miserable failure. American helicopters collided in the desert staging area, resulting in a loss of life, and the mission was aborted. It was stunning news.

At the same time, I received word that the International Molders Union was passing out cards for our associates to sign at the plant entrances. The purpose was to force an employee election that they hoped would lead to the company being unionized.

This was disturbing for a couple of reasons. First, we prided ourselves on being a progressive employer who was sensitive to associates' needs. Second, as a sole supplier to virtually everyone with whom we did business, it was well understood that customers would be gravely concerned by the potential for work stoppages that came with unionization. In the event we were organized, there was no doubt they would protect themselves by attempting to develop alternative sources of supply.

USPL technology changes, growth, and the addition of a new factory requiring some associates to work at a different location caused uncertainty for a few, dislocation for a few and, in some cases, inconsistency in management practices. As a result, enough cards were signed for the union to be granted an election. A campaign followed, during which I had numerous meetings with all associates to tell them why we believed representation was not necessary and about the potential for business loss it would create due to the company's sole supply position.

On June 19, the National Labor Relations Board conducted the election in both factories. The Molders Union lost in an overwhelming vote of 87% to 13%. Almost always, such votes are much closer because over half the eligible employees have to sign cards petitioning for the referendum if it is to be held. Although the union had secretly been conducting the sign-up campaign for some time, we were embarrassed that they could obtain enough signatures. When our side of the story was told, and everyone learned the facts about the Molders historical questionable behavior as well as their record in representing employees at other companies, it quickly became obvious we would win. The margin was gratifying and sent a message to others on how our associates felt about their company. The fact there even would be an election sent a reminder message to those of us in management that it was critical to have excellent two-way communication with the work force and consistent application of employment policies.

Calendar 1980 was again economically tumultuous. The U.S. inflation rate was an incredible 13.5%. Bank prime lending rate ranged from a low of 11.1% to an astonishing high of 20.4%. The Japanese yen varied from a high of 203 to the dollar to a low of 261. This meant that from the high to the low, our lenses cost the Japanese TV makers 29% more because they were buying with dollars. The environment of huge material and wage cost increases made it a difficult time to accurately build and execute business plans.

Fiscal year 1980, ending June 30, resulted in sales of $9,363,000, up only 6%. Net after-tax profit was $729,000, down 28% due to an aggressive expensing policy on massive-optic machinery building and the move to the new factory. The economic turbulence was also taking its toll on operations.

This was the peak year for LED semiconductor sales, which increased slightly to $1,869,000. Unfortunately, it was a business that would soon drop off significantly as calculator makers began shifting away from LED's to liquid crystal displays. Polaroid was now an insignificant customer. In addition, the component and system category slipped 32% to $570,000; Honeywell auto-focus modules, although still over $1,000,000, were down 26%. In just the fourth year as a category, projection video optics provided sales of $4,776,000, up 44%, now representing slightly over half of the total volume. At this point, we had not yet begun to ship the Delta II-M to Sony for their coffee table model that was introduced in September.

For the calendar year, projection televison sales of high quality three-tube systems were up 90% to 57,000 units. Sony's new set substantially boosted these numbers.

In the last part of the 1970's, personal computers were being introduced to consumers principally by Apple. At the Consumer Electronics Show in Las Vegas in 1978, we saw a large display by Apple, and wondered why anybody would want one of these things. In 1979 I was the education chairman for the Cincinnati Chapter of the Young

Presidents' Organization, and brought in a Futurist from the University of California—Davis to tell us what the world might be like in the years ahead. In an all-afternoon session with roughly forty in attendance, his biggest point was that the personal computer was going to become a very important part of our lives and that most of us would eventually have them. At the break, I was strongly admonished by unhappy members for wasting their time by bringing in a speaker with such a nutty prediction. So much for the clairvoyance of the Young Presidents!

On a September 1980 trip to California's Silicon Valley, I spent the flight time building a spreadsheet analyzing prospects for the upcoming three years. It included a detailed breakdown of sales, expenses and capital requirements. This type of exercise was done often as conditions changed. It required a great deal of thought on what we believed with respect to every item. That afternoon I called on Dave Hillman, an old Litronix friend who was then working for Jack Magarian at National Semiconductor. After business was discussed, he said he wanted to show me something at a nearby store. The shop we visited sold only personal computers and software. It was probably one of the first of its kind. Dave sat down at an Apple II and demonstrated what could be done with the first spreadsheet program called VisiCalc that had just come to market. Having worked hours earlier in the day to construct one manually, I was amazed with the different scenarios that could be run in a matter of minutes. He inputted a fictitious business plan and manipulated it in a multitude of ways.

On returning to Cincinnati I enthusiastically told our management group about what could be done on this small machine with the program. Soon thereafter, an Apple II, the VisiCalc software and a printer were purchased for $7,000. Len Kosharek had worked with the computer before joining us, but not this particular program. He became the company's personal computing expert, and soon it was being used for business modeling, and scheduling as a myriad of other applications we only dreamed of doing on our IBM mini-computer that was used for accounting tasks. From this beginning, personal computers would proliferate in USPL to the point of being used in every facet of the company's activity. Although the benefits of the computers were enormous, I never felt my input numbers were thought through as well as when the process was done manually. In any event, the Futurist from California had it right.

For ten years I watched a neighbor, William F. Mericle, build the plastic injection molding machine business of Cincinnati Milacron from a clean sheet of paper to the largest maker in the world and the most important division of America's biggest machine tool company. It was an impressive achievement admired by all who knew about it. Over the years I talked to Bill frequently about what each of our companies were doing, and sought his advice on many occasions. We had a good board of directors—but no one on it had manufacturing experience, even though USPL was a manufacturing company. The other members agreed it was a deficiency and that Bill Mericle would be an excellent addition if I could convince him to give us the time. Fortunately, he was interested in what we were doing. Bill joined the board on October 15, 1980, and just as we had hoped, was a superb contributor filling a void in the group's set of skills.

On November 4, Ronald Reagan was elected President of the United States. That evening I took the new Sony coffee table projector and screen to an election party. Everyone there was amazed by the excellent picture quality, and wanted to know about price and availability. It was an encouraging sign for our future.

In 1980 we had in excess of fifty visits with eighteen major television manufacturers—another encouraging sign.

# 39 PTV Lens Proliferation and Competitive Threat

By 1981, both our television optical design and massive-optic manu-facturing learning curves were steep. Substantial development money was being spent in all areas as we raced to stay ahead of the Japanese lensmakers and satisfy the numerous requests for special features that were coming in from customers. All this effort resulted in a prolifera-tion of over ten Delta lenses, each available with variations allowing customers to use a version compatible with their system configura-tions. Already, it cost well over $100,000 to prototype and production tool a lens—a number in those days that made us think hard to be sure the right bets were being made.

Jack Welch, the legendary CEO of General Electric, has been quoted as saying, "When the rate of change on the outside exceeds the rate of change on the inside, the end is in sight." Everyone in a key position clearly understood that USPL was considered a threat by the Japanese lensmakers, who were feverishly working to eliminate the need for our existence. The importance of Welch's point was keenly felt by Brian, Dave, me and others in our leadership group. For the next few years, every trip to Japan would involve the playing of a little game we labeled "Gotcha!" At some point in meetings, a few of the customers would tell of some new desirable development the Japanese lensmakers were working on or about to demonstrate to counter what we were doing. Although they were always very nice in the way the information was imparted, we nevertheless took it as, and internally called it, a "Gotcha!" With the large stream of new designs and fea-tures coming from our product development activity, we were always be able to respond by revealing something new coming from USPL that represented important advancement. Among ourselves we called it *our* "Gotcha!" I cannot remember a time when they put forth a "Gotcha!" that we couldn't counter with a better one. By 1983 or 1984, the game was no longer played.

In Japan, by mid-1981 there was an intense interest in plastic lensmaking. We were contacted by one of their trade organization's Washington office, requesting that a plastic optics study consortium of roughly thirty people from various optic companies visit USPL. We told them as nicely as possible that virtually everything in our factory was of a proprietary nature; therefore, there was little that could be seen. The last thing in the world we wanted to do was teach them how to compete with us. Our preoccupation and penchant for not reveal-ing manufacturing technology was well known to everyone who had any knowledge of USPL. Nevertheless, they pressed for the visit again, indicating their interest to see us on a swing which included tours at Kodak and Polaroid that had already been arranged. Finally, I told them that from their standpoint the meeting would be very unproductive and that in addition, the timing was not convenient due to scheduling conflicts.

The results of the consortium's study were reported during a Tokyo conference in February 1982. One of our customers gave me a copy of a 100-page book the group issued at that time. The bulk of it was filled with pictures and text from USPL brochures, articles and technical papers. Promotional pieces produced from earlier times by American Optical and Bell & Howell, were also included. My 1978 OPTICAL SPECTRA article, entitled "Big Optics for Big Screen Tele-vision," and Brian's and my 1979 IEEE paper were reproduced in their entirety. It gave us an eerie feeling to see our own pictures, photos from USPL brochures, and the text of things we had written repro-duced in a book surrounded with copy in the Japanese language.

There was no doubt they properly wished to expand their knowledge of plastic optics—no doubt in our minds that the knowl-edge would be used to obviate the need for the world's television man-ufacturers to do business with USPL. Our policy of keeping manufacturing technology a secret had again served us well.

Projection television makers were continually working to improve the brightness of their systems. This was accomplished primarily by developing picture tubes, or CRTs, that could be driven harder with additional electrical current. As this occurred, more CRT faceplate heat was generated which was, of course, in close proximity to the lenses. The elevated heat was beginning to expose a problem with our all-plastic Delta lens designs. As the large power element increased in temperature it expanded, causing the curve to change and the entire lens to defocus slightly. It was a problem the glass makers didn't have because plastic expands with heat ten times that of glass. We recognized this limitation as serious and fundamental.

In the years ahead, there would be many approaches to overcome the problem. Sony, and eventually others, would put liquid cooling chambers on the front of the CRTs to drain off excess heat. Brian and Opcon worked to develop designs that altered negative and positive power with the objective of canceling out expansion changes. This was only partially successful. By the end of 1981, we concluded the ideal way to correct the deficiency was to put as much power as possible in a spherical glass element and use plastic elements for gaining the performance advantages that can be realized with aspheres. Such lenses were called hybrids. Japanese optic makers were still producing expensive all-glass, six and seven-element designs. We were making three and four-element, all-plastic designs, except for institutional versions that would have five or six elements. At this point they didn't know how to make plastic aspheres, and we had no capability to manufacture spherical glass elements.

We would struggle for a couple of years with this limitation. Most of the U.S. glass-making business had been lost to Europe and Japan, but a few companies remained that made military optics, overhead projector lenses and lenses for miscellaneous low quantity applications. Brian and Opcon had suitable designs that allowed us to obtain outside glass element quotations, but they were not attractive and would have greatly increased Delta prices. We knew that if a good answer was not found for the thermal expansion problem, our television lens business was excessively vulnerable to competition.

A second problem with the all-plastic designs was beginning to be brought to our attention by institutional PTV customers. It was called color fringing or, more properly, chromatic aberration. The condition was more prevalent in the longer focal length lenses. What the viewer saw was a minute color band on the edges of objects that had sharp lines. It occurred because the different wavelengths of red, green and blue light focus at slightly different points. This condition could be eliminated with color-corrected designs using positive and negative power with different refractive indexes and dispersion qualities. This was a problem because our lenses were made of acrylic and, although styrene had the right characteristics to be combined with acrylic to achieve color-correction, we didn't know how to process it into acceptable elements. The glass lensmakers had a much wider choice of raw materials to eliminate chromatic aberrations. For us the solution was to build glass/plastic hybrids, but it would be costly and we still didn't have a good source for glass elements. For the Japanese glass lensmaker the solution was easy but even more costly. Now, in addition to the heat problem, there was a second reason to find a source for glass elements that would allow USPL to build hybrids.

Men loved projection TV sets, but most women hated them because they took up too much room space. We continued to search for ways to reduce cabinet size. In May of 1981, Zenith introduced a unique consumer model that looked like a normal sideboard cabinet when not in use. When the remote control "on" button was pressed, the top folded back and a 45-inch screen was elevated by a motor with a picture projected to a screen from unseen Delta IVs in the rear. We had been working on it with them for a year, and it was awaited with great anticipation. A Zenith executive invited me to their Las Vegas dealer meetings where it was first revealed. With great fanfare, it mysteriously appeared on a huge stage. Our friend Bruce Huber, the vice-president of marketing, started talking about the beauty of the object and then the need for a truly unique new consumer electronic product. None of the dealers in the packed auditorium had any idea what it was. Then, with more spotlights focused on the cabinet and a blast of introductory music, he pressed the remote button, the top folded back and the screen magically appeared. The dealers gave Bruce a wildly cheering standing ovation.

In the market, the Zenith set was moderately successful at a staggering list price of $3,750. It was still slightly over 29 inches from the front to the back, which a number of the potential buyers found to be too deep. In any event, it was an innovative attempt to solve the configuration problem and resulted in much favorable publicity for Zenith.

Near the end of summer in 1980, Brian came up with a marvelous idea about how to reduce PTV set size. The in-line CRT lens combination all makers were using determined cabinet depth and height. Delta IVs were 8.5-inches long, and the CRTs were about 11.5-inches long. This meant an assembly of over 20-inches had to be fitted into the cabinet. His creation was a design with a housing that incorporated a mirror which turned the image ninety degrees between elements in the lens. This ingenious technique allowed the long CRTs and the electronics attached to their end, to be tucked up along the back side of the cabinet, this dramatically reduced both the depth and height. It would mean that Magnavox and Sylvania models' depth could actually be cut in half.

To prototype the design for feasibility and demonstration, we determined that a complicated injection molded housing tool would have to be designed and built on a tight schedule. Jim Ballmer and a team of his engineers took on the project and gave us an extremely long lead time for completion that would result in a year of market introduction delay. This was because of the way our prototype delivery would have fit with the manufacturer's product design cycles which, in turn, drove their new product introductions. We told Jim the lead time was unacceptable and that we had to demonstrate the lens in half the time he was promising. I dubbed it "Project Impossible" and challenged him to use whatever resources were needed to get the job done. Jim came back to say there was no way to get the plastic housing mold built in the time frame we wanted. He proposed that we make the prototypes with cast aluminum housings. This more than cut the development time in half. On January 22, 1981, we successfully demonstrated the lens and celebrated with a "Project Impossible" party for the team that worked day and night to rush it to completion.

The new lens was named Compact Delta I. The concept was an instant hit as we showed it to all our existing customers throughout the spring. It was not revealed to Mitsubishi, a leading Japanese maker that used all-glass lenses, because we believed they were so wedded to their supplier at that time that the information would lead to a search for an all-glass alternative. At a Sony luncheon meeting in Tokyo, with the Senior Managing Director in charge of all consumer video products, Masaaki Morita—brother of co-founder Akio Morita, Brian sketched the Compact Delta I layout. Before he could finish, Mr. Morita uncharacteristically exclaimed "fantastic idea!" Reactions everywhere were similar.

Samples of the lenses in cast aluminum housings were priced at $8,500 for a set of three, which would allow the building of one prototype set. The cost was high because the various metal parts had to be cast, machined and anodized. With a big demand, the price included a healthy markup to partially recover the development expenditures.

As Compact Delta I prototypes were being shipped, Sony had been selling the coffee table projector only since the previous fall. North American Philips was about to introduce Magnavox and - Sylvania brand sets using Delta IV. Zenith was offering its pop-up version, and RCA had just revealed a one-piece mirror foldout model with Delta IVs made for them by Hitachi. Hitachi was also offering a similar model under its own name. It was a problem for them to consider a totally new design with Compact Delta I before having feedback on models that were just reaching the market.

On March 9, 1981, *The Wall Street Journal* carried a positive story entitled "Makers Bet Millions That Big TV Screens Will Be The Next Rage In Home Entertainment." A paragraph in the article, with the heading "New Plastic Lens," read as follows:

> Until recently, the relatively poor quality of big screen, or "projection," TV discouraged most manufacturers. But their interest revived when they noticed a marked improvement in images televised on sets being sold in Japan. As it turned out, much of the improvement came from a new plastic lens manufactured by a small company in Cincinnati, U.S. Precision Lens. The company supplies such Japanese electronic giants as Sony, Toshiba and Hitachi. RCA, among others, credits U.S. Precision Lens with the quality advances that allow RCA to put its own name on big screen TV.

Unfortunately, the heat problem that was experienced with Delta II-M and Delta IV was also present in Compact Delta I. To minimize it, manufacturers focused the lenses at the factory under elevated temperature conditions. This resulted in a slight out-of-focus condition when cool sets were first turned on. It became better when the heat reached maximum levels where the factory adjustment was made. The makers lived with the problem because there was no good alternative. At the time the available glass systems had focal lengths that pushed the sets to even larger sizes, and they were considerably more expensive. Furthermore, there was not enough large glass lens-making capacity in the world to handle the existing demand if they wanted to go to all-glass designs.

In 1981, projection television sets were selling from $2,500 to $4,200. We quoted the lenses in kits of three assemblies or what was required to build one TV. In 5,000 to 10,000 quantities, our approximate three-lens kit price for Delta II-Ms was $145, Delta IVs $170, and the forecasted 1982 price for Compact Delta I was $185. The first models incorporating Compact Delta I were introduced in 40-inch Magnavox and Sylvania models at the 1982

Summer Consumer Electronics Show in Chicago. These were soon after followed by 40-inch and 45-inch offerings from Hitachi, RCA, Sony, G.E. and Zenith.

# 40

### See It—Smell It—Taste It—Smoke It:
### Understanding Our Japanese Customers

After the 1980 union organization attempt, we re-examined our personnel policies and how they were being executed. Moves were made to ensure consistency. To improve communication with associates, a company publication was initiated that would be produced periodically. Edited by human resources specialist Kathy Sutton and called *FOCUS*, it carried articles on business activity, plans for the various segments, plus overview commentaries by Dave, me and others on items affecting USPL generally. It was an effort to augment what we called the "company meetings" where every few months I and sometimes others would address all associates in a series of sessions on important issues. After the formal presentation, those meetings were used to answer questions on whatever was on anybody's mind.

I sensed the company meetings were beginning to take on a sameness as we frequently talked about the high Japanese quality standards, requests of customers and concerns about competition. I could imagine people saying to themselves, "Here it comes—he's going to tell us how demanding the Japanese are again." For all the talk, the fact was that most associates had no idea what Japan was like or who the faceless names were they kept hearing about that had so much influence over our future.

In a departure from the previous routine of providing education and understanding about our business in that part of the world, I concluded something very different needed to be done. When planning the next trip to Japan, we decided to take extensive slide pictures of the companies and people we were dealing with. The idea was to create a photo illusion for our associates that they accompanied us on the trip. Brian helped take pictures of laboratories, typical meetings in progress, key players, competitive lenses and even the lunches that were served in customer dining rooms. The trip photos started at

Tokyo's Narita Airport, included our hotels, the local as well as bullet trains used for travel, ancient temples in the parks on the weekends, and customer dinners in Japanese restaurants—with us awkwardly sitting on the floor eating food with chopsticks. Much of it looked quite strange by American standards. The Japanese customers knew what we were doing, thought it was a great idea, and cooperated with the photo sessions, in which they occasionally waved to the camera.

At Narita Airport just before the return trip, we bought bags of very different tasting popular Japanese candies, cookies, cigarettes (in those days many associates still smoked), and a large quantity of a unique, stringy, dried fish snack food that carried with it a strong unfamiliar odor.

At the next company meeting, I told our associates this was going to be different. They were about to take a fantasy trip with us to Japan and meet many of the people we depended on to buy the lenses that kept us working. I added that not only were they going to see and hear about these customers and their country, but we also were going to get to everyone's senses with the sweets, delicacies and smokes of the nation as they watched. Trays were passed with the goods for consumption during the presentation and with their families later.

The meeting was a big hit, and accomplished what we had hoped. Our associates had a much better understanding of the Japanese we were doing business with and the country in which they lived. For weeks thereafter, people would stop me in the halls or the factories to comment about it, ask a question, or tell me their kids couldn't believe anybody would eat the stringy, dried fish, or their husbands thought the cigarettes were really different.

It was a fun thing to do, and simply represented another one of the many ways we attempted to engage our workforce in the dynamics of their company.

# 41 A Mistake

When asked for the reasons as to why USPL was successful, one of the claims I would make, until 1981, was that we had not made a major mistake—or at least not a major mistake we knew about. It was never said in a boastful way, but rather as a recognition of good fortune.

In the fall of 1980, Mr. Yohsuke Naruse, of Sony, requested a meeting during a Tokyo visit to discuss a new plastic lens project. Naruse was the manager of optics in the disc development division. We were asked to prototype a very small aspheric digital pickup lens for a video disc player. It was to be part of the optical system that read, or picked up, the information that had been recorded on the discs allowing video images and sound to be processed. Everyone recognized that video discs provided significantly superior picture resolution compared to video tape. Although they couldn't be used for home recording, some observers thought the systems would become quite popular. We thought it was a limited market, but since the Sony relationship had become so important, we were cooperative.

Soon thereafter we learned Naruse's real interest was in pickup lenses for compact digital audio discs, which had recently been introduced to the market by Philips and Sony. The potential for these devices was much larger, although it was unclear just how important they would become. Advent president Bernie Mitchell, had been president of Japan's highly successful Pioneer Corporation of America and built it into a leading audio components marketer. I asked him what the future prospects were for compact audio discs because he was considered an expert in the field. He said that since audio tape was compact, well established in the market and could also be used for recording, he thought Philips and Sony would not be very successful. We didn't know who was right. Bernie had a lot of experience in the field; however Sony, with its spectacular track record, was investing heavily in the technology.

It was not a good time for USPL to take on a time-consuming project, because the product development demands of the television business were intense and had our people stretched thin. Nevertheless, Dave Hinchman quoted a three-phase program on a "best effort basis" that was expected to cost over $60,000. Sony paid a $10,000 deposit for the first phase.

The lens was tiny and had to be of extremely high quality. Opcon performed design studies so the level of accuracy needed could be better understood. We built a mold insert and dedicated a small molding machine to the effort. The first lenses were not good, but with succeeding attempts they got better. Initially Naruse was encouraged, but as the project dragged on we thought he was becoming apprehensive. I was becoming apprehensive, too. The effort was taking a lot of time and resources, and we came to realize that a higher precision, small specialized molding machine would be required to increase the chances of success, and that better testing equipment was needed. In addition, these lenses would also require sophisticated coatings that at the time were beyond our capability.

We had a dilemma. Either we had to dedicate more capital and talent to the project, or it would fail. Of course, there was still the question of how big the market would be if there was success—and it was clear the cost would far exceed $60,000. Most of all, we were concerned that ours was not really a "best effort," and unless there was a change, it might damage the Sony relationship.

Remembering how LED bubble lens customers who greatly respected USPL's unique ability in the early 1970s were disappointed with our performance on the less-important reflector digits and how it made them step back to wonder if they had correctly assessed our overall capability, I feared the same thing might happen at Sony as a result of this project. We decided to talk frankly with Naruse about the status of the effort, what needed to be committed to it to increase the chances for success, and our concern about being able to provide the resources to do it in a timely manner. In a meeting at the New Otani Hotel I told him we were deeply concerned that this might lead to bad feelings at Sony and offered the option of ceasing to work on the project, not billing for work beyond the deposit, and returning the $10,000 that they had previously sent to us. We didn't need to do this because as a "best effort" development, contract success was not guaranteed.

Mr. Naruse thanked us for being so forthright and said he would think about it. A couple of days later we met again at the New Otani, where he agreed to end the project and stated they would accept the return of the deposit. Because of the way it was handled this contract never became a problem in relations with Sony, but with perfect hindsight, I had made a mistake.

Compact disc players would soon become a hugely successful consumer product, surpassing tape players as the preferred technique to deliver superior sounding music. Konica Corporation successfully produced the lenses, and for a few years supplied the CD player makers enormous quantities at good prices. Eventually there was competition and prices dropped to commodity levels, but it was excellent while it lasted.

Bernie Mitchell was wrong and we were wrong. If we had put the resources to the project that were required, there would have been success. It wouldn't have been anything approaching that of the television lens business, but it would have been significant. Never again would I say that we had not made a major mistake.

# 42 An Innovator Fades Away: The Demise of Advent

It was sad to watch; Advent, the company that pioneered consumer PTV and brought us into a category that far surpassed all our other markets, continued to flounder. Peter Sprague's 20% interest and voting control agreement were causing him grief. Pierre Lamond, the man with such a good track record at National Semiconductor, and whom he had installed as president, failed to improve things and was fired in May of 1977, after only fourteen months. Sprague took over as president in 1978 and, after years of losses, the company actually earned $2.3 million on $36 million in sales. We suspected those profits were enhanced by generous accounting assumptions and dealer pipeline filling for the Model 710. In 1979 the company was back in the red with a loss of $2.6 million. Sprague bowed out as president, installing William Anderson, a former RCA and Sharp Electronics marketing specialist.

Bernie Mitchell, who as president of Pioneer Corporation of America saw it go from $2 million to $230 million in sales, was out of a job. He sent a letter to Sprague telling him Advent would soon be out of business if it stayed on its present course. He suggested a cure which involved buying components from Japanese makers such as video disc players, cassette players, amplifiers and VCRs that would be sold under the Advent name. His idea was that home entertainment centers, which included a PTV and all these devices, would bring a revolution to the audio-video industry. He suggested this $36 million company could be a $156 million company by 1985. Sprague took the bait and hired Mitchell as vice-chairman and CEO. Anderson continued as president, but reported to him. A $3.2 million cash infusion didn't last long, and Sprague was horrified only a few weeks thereafter to learn Mitchell needed more.

We didn't know all these details at the time, but there was a high level of concern at USPL over Advent's ability to survive. They were still buying the original David Gray f/1.4 design, even though our Delta f/1s were obviously superior in sharpness and brightness. No improvements were being made in their CRTs or other electrical components. While nothing technically was happening at Advent, the other makers were aggressively working to upgrade all aspects of their systems.

On November 21, 1980, Bernie Mitchell wrote to me with concerns about our ability to supply their needs. He was forecasting 13,400 sets for 1981, and 22,800 for 1982, with part of the second year's volume going to a new factory they intended to build on the West Coast. Four days later I wrote to him, pointing out that we would have no problems fulfilling their needs if given reasonable lead time. We had been pressing Advent on their dragging receivables, and my letter stated that from that point on we would not ship if payments were beyond the normal thirty-day terms.

Bernie had an idea for a PTV that he thought would solve their sales problems. It was configured as a table with a top that folded back to become a 50-inch screen for a one-piece projector. Although he was very excited about it, we had doubts as to whether it would be attractive to buyers. In addition, with no optical or electronic improvements, picture quality suffered in comparison to the competition. At the January 1981 Las Vegas Winter Consumer Electronics Show, the Model VBT-100 projector was introduced with all the flashy hype Mitchell could muster. We were surprised and encouraged when he told us he had written orders for over 3,000 sets at the show.

Advent had paid its old invoices to within our terms, and we expedited production to meet the demand for the new orders booked at the show. To our surprise and shock, on March 17 they filed for bankruptcy under Chapter 11. We were the largest trade creditor with $300,000 in receivables, all of which were within thirty days on the date of the filing.

I flew to Boston and found a bankruptcy lawyer to represent us. Thereafter there was a meeting with five or six of the other large creditors, at which our lawyer was chosen to represent the new creditors' committee. I declined to be chairman because of our Cincinnati location, and pushed for one of the Boston-based company presidents to take on the job. He did, and together we all received an education on bankruptcy. Unfortunately, by the time our education was complete, it was apparent we had done the wrong thing. Under the court's auspices, Mitchell and others kept the company going as they sought capital infusions and a buyer for the entire enterprise. We were continually being presented with the hope of a rescuer appearing as the assets of Advent were consumed. In hindsight, what we should have done was demand everything be shut down and liquidated. If that had happened, we might have received twenty-five cents on the dollar and saved a lot of time. As it was, we received nothing. The only good outcome was our lawyer being chosen to represent the committee. That meant the assets of Advent were used instead of USPL's money to pay his fees.

It had been seven and a half years since I made what was my most important sales call ever on Henry Kloss. We were now selling television lenses at a dollar volume level of over ten times what the entire company's sales were then, and in the next year they were expected to more than double. The company that brought us into the industry and should have been a leading maker was now gone due to self-inflicted wounds. Although not happy about it, for us $300,000 was not very much to pay for the PTV opportunity that came our way as a result of doing business with Advent.

# 43 1981

Every major television maker in the Free World had a development team and plan in place to enter the projection TV market. Advent, G.E. and Sony had done the pioneering work, laying the foundation that was attracting all the others to the product category. Mitsubishi, Panasonic, Quasar and Kloss had delivered new models the previous year. Kloss used USPL-made corrector plates in its reflective system to eliminate spherical aberrations, giving us a sales content of about $14 per set. The Mitsubishi, Panasonic and Quasar models contained Japanese glass lenses. 1981 brought introductions by RCA, Zenith, Magnavox, Sylvania, Hitachi and Pioneer in the U.S., plus Grundig and Philips in Europe—all with USPL optics.

Our forecast showed volume would double starting in the second half of the year, which was the first half of fiscal 1982. To be ready for it more space was needed, along with a substantial amount of new production equipment. A $2 million tax-free industrial revenue bond was purchased by the First National Bank with floating terms at 60% of the prime commercial lending rate. Equipment builds were started in all areas where capacity was needed, and a 20,000 square-foot structure was added to the McMann headquarters plant.

In an April board meeting, I reviewed six priorities for the year:

1. Aggressively continue exploiting the PTV opportunity every way possible.

2. Support Standard Optics as we look for significant new product opportunities.

3. Maintain and improve excellent workforce relations during the anticipated period of great change. Preserve non-union status. (We then had 425 associates.)

4. Continue to make meaningful improvements to the massive-optics technology.

5. Employ a highly qualified chief financial officer.

6. Do a preliminary investigation on the advantages and disadvantages of a public offering of our stock.

USPL did not have financial management leadership other than what Dave and I provided, which was largely in the area of planning and budgeting. Instead we had an accounting department that kept track of what was being done and the assets. In the past this was not a concern because the company was small and not financially complex. Also, annual audits by Arthur Andersen provided comfort on accuracy. But now, given our size and anticipated rapid growth, it was becoming clear the old ways of viewing finance would not be acceptable for the future. We needed a financial leader who would make the function proactive in providing options, predicting outcomes and generating analysis that allowed us to better understand the consequences of actions. We particularly needed improvement in the accurate calculation of product costs.

The task of finding a chief financial officer was assigned to Jerry Behne. We hoped to find the person in the Cincinnati area. Most of the potential candidates were in public accounting firms, but we made a decision that extensive financial management experience in manufacturing environments was a prerequisite. This eliminated all the public accountants. After months of searching and a number of interviews, Jerry concluded we had to look outside of Cincinnati. *Wall Street Journal* advertisements were run in a few large cities which brought many responses. Jerry studied them in detail and pared the list down to the top choices.

One resume from Chicago jumped out at him as being significantly better than the others. It was from a thirty-seven-year-old fellow named John C. Collins. He had remarkable education and experience. On paper, he appeared to be just what we were looking for. Jack grew up in the Boston area and went to the Massachusetts Institute of Technology, where he received a B.S. degree in industrial management, with concentration in finance, accounting and electrical engineering. After that, he was employed in Elgin, Illinois by a maker of molded rubber products as a production engineer and eventually as the chief pilot engineer. His next job was with a metal products manufacturer as manager of accounting and then as director of engineering services, which involved managing a staff of twenty-eight in manufacturing engineering and cost-accounting functions.

During his time with this company, he earned a masters of business administration degree from the University of Chicago in 1972 and became a certified public accountant in 1973. In 1978 he was hired as controller for a multi-national electrical products maker. Every line of his resume indicated experience that was relevant to our needs, and his education obviously could not have been better. Jerry visited him in Chicago and arranged for Dave and me to meet him there at the Summer Consumer Electronics Show.

The show was hectic with customers coming and going at our suite in the Ambassador East Hotel. Under conditions that were not ideal because of everything else we had on our minds, an interview was squeezed into a break one afternoon. For some reason still unknown to me today, I was not overly impressed by him in this meeting. Jerry was bewildered by my reaction, and uncharacteristically nearly demanded that we do everything possible to hire him. Having a deep respect for Jerry's judgement, we followed his advice. Late in the summer, Jack agreed to our offer and joined USPL on October 5.

It was obvious immediately that Jack was a perfect answer to our needs. Swiftly, he reconfigured the accounting system and soon was delivering more timely financial reports along with providing the proactive data that had been missing. He quickly gained the confidence of all the key managers who began to come to him not only for financial information but also for what was recognized as sound judgement. With an unflappable and mature manner, he filled a void superbly .

The timing of his arrival could not have been better. In less than two years, events would unfold in which Jack would make contributions far greater than anything we had envisioned. Beyond that, he became a highly valued adviser to me on a wide range of issues. Jack Collins was an extremely important participant in the building of USPL. In 1996 he joined the R. A. Jones Company where he made similar significant contributions as its chief financial officer.

The huge forecasted growth in the PTV business was causing some of the largest buyers to be nervous about our ability to meet their needs, and their dependence on one manufacturing location. It was the same concern the California LED makers had expressed years earlier. They were pressing us to do something to make them more comfortable. After much thought and analysis, we decided to build another TV lens factory far enough away that the associates working in it would not be considered in the same labor pool if an organization attempt was made. Beyond that requirement, we wanted it to be as close to the headquarters location as possible to minimize the difficulty in moving materials back and forth, as well as making it easy for management to visit.

With the help of the Miller-Valentine Company that had built our Bach-Buxton plant, we found a few acres of land on the Cincinnati circumferential highway I-275 near the Cincinnati airport that was ideal, except for the fact there was no exit to the road that crossed at that location. The Commerce Department of the State of Kentucky was so interested in having USPL build there that it sent a helicopter to fly Jack Collins and me to the State Capital in Frankfort for negotiations. This resulted in our agreeing to buy the land and build a 63,000 square-foot factory if the state would build entrance and exit ramps to the overpass road that abutted our property. They agreed, and the deal was struck. We obtained a commitment from the First National Bank for another $5 million industrial revenue bond financing and began waiting for Kentucky to install the ramps. It was a process that would ultimately end with an ironic twist.

There was a need to add technical depth to our research and engineering area. Serendipitously, we received a resume from Dr. James Bohache, a Cincinnati native who had just received his Ph.D. in optics from the University of Rochester, one of the two leading optical schools in the United States. Jim's particular interests were metrology or optical measurement and lens fabrication. We had an immediate need for the first skill and suspected the second would become important in the future. He joined us in August, reporting to Brian Welham. Jim was the first Ph.D. to be directly hired by USPL and became an important contributor.

One of our European customers was buying Delta lenses; Thomson-Brandt, France's largest electronics company and second in size only to Philips on the continent. We had made a number of visits to their factory in Angers. It was always an enjoyable trip from Paris through the beautiful French countryside on what may have been the finest train in the world at the time. The man in charge was a pleasant fellow by the name of Alain Granger. On one occasion Joyce and I had a delightful dinner with Alain and his wife, Monique. We learned he would soon be coming to the United States, so I invited him to visit USPL, stay at our home and allow us to entertain him in Cincinnati. He accepted.

After meetings at USPL and a factory tour, Dave, Brian and I took Alain to the Maisonette French Restaurant—the longest continual Mobile five-star gourmet award winner in the country. When making the reservation we told the maitre d' about our special guest and asked if the chef, also a native of France, might come out and say a few words to him. To my surprise and bewilderment, Alain was critical of the Maisonette almost from the moment we arrived to the time we left. In spite of having the maitre d' and chef fawning over him, he kept finding fault—from serving the salad in the American custom at the beginning of the meal, to a misspelling of a French word in the menu. The wine list didn't impress him, either. All of this was beginning to irritate me as the evening progressed, but I didn't indicate displeasure.

The next morning, still slightly irritated, it became clear to me what we would do. When lunchtime arrived, I told Alain that last night we attempted to offer him the best we had in his country's style of cuisine. Today we were taking him to a type of restaurant that was unique to our area—a Cincinnati chili parlor. Cincinnati's chili is a combination of chili sauce, beans, onions and grated cheese served on a bed of spaghetti. It was a hot day, and the place we visited had an air conditioning problem which required the windows to be opened. Also, it appeared to be short of help because the tables were not very clean and the wait was longer than usual. As Alain sat there watching outdoor workers coming to get takeout orders and watching the flies buzz by, one could only imagine what he was thinking. Finally the food was served and Alain, after a long pause, looked up at me and said, "Roger, actually that was a very good restaurant we went to last night." We all did everything in our power not to embarrass him or ourselves by containing our urge to break out in laughter. I liked Alain Granger, and Joyce even liked him better. Shortly after he departed Cincinnati she received a dozen beautiful roses as a thank-you for having him stay at our home, and reminded me it had been some time since her husband had sent flowers!

In an article for our company publication, *FOCUS*, titled "Report from Tokyo," Dave Hinchman talked about performance, things on the Japanese customers' minds and things on our minds.

> Brian Welham and I just returned from a successful visit with our Japanese customers. Although most of our meetings were very technical discussions about projection television lenses, several conclusions about this trip are important to all of us.
>
> First, I cannot emphasize enough how important it was that we met or exceeded our Delta lens production goals for three months in a row. All of our customers, especially Sony and Hitachi, have been worried

about our ability to expand fast enough to keep pace with their need
for lenses. We have a very ambitious production ramp, and the fact
that 'we did it' both impressed and reassured them. Thanks to all of
you who worked so hard to pull this off.

Second, the quality of our lenses continues to be excellent.

Third, they were reassured by our decision to proceed with a new plant
to produce television projection lenses.

Fourth, Japanese lensmakers continue to work hard to develop plastic
lenses to compete with us. We have been told that there are over fifty
companies in Japan working to make plastic lenses. We know about
eight of these for sure. However, because we are developing exciting
new products, and because we are getting the job done on quality and
production, we are making it very difficult for this competition to gain
on us. Although I expect the competitive race to be tough in the future,
I believe that if we continue to work together as a team like we have
these past months, we will win out.

The April board meeting's sixth priority for the year was to
preliminarily study the idea of making an initial public offering (IPO)
of our stock. There was no urgency on the question. The purpose was
simply to better understand it as a future option, and to have the home-
work done if it was determined to be a desirable thing to do. Selling
shares to people outside the company was something we did not want
to do unless there was a compelling reason. Such reasons would be
the need for capital, shareholders' need for cash, or a shareholder's
need to diversify assets. We had been easily financing growth through
retained earnings and the issuance of industrial revenue bonds. With
the prospect of even faster growth of PTV, we could envision a day
when an equity infusion might be required. I did not feel that the
Howe family, with a substantial majority of the shares, needed to
diversify its holdings yet, and none of us had a personal cash demand.

In the past, various venture capital funds sought to invest
in USPL. The talks only became serious with Greylock, a fine Boston

firm with the reputation of being a patient and helpful investor. A deal was never consummated because neither the company nor the shareholders needed the money. Starting in 1978, a number of companies inquired directly or through intermediaries if USPL would be interested in being acquired. The answer was always no.

The IPO idea was investigated with McDonald & Company, Shearson American Express, Rothchild, Unterberg & Tobin, and Oppenheimer. My conclusions were that it would be fairly expensive. USPL had the potential to sell at an excellent price earnings ratio because of explosive growth, but the potential volatility of the projection television market could disappoint shareholders. Growth consistency is highly prized by investors; even though a company has a fast growth rate over time, there is a substantial market price penalty for an occasional down year.

USPL had another reason not to be a public company. Many of our consumer electronic customers were not profitable in the notoriously competitive television business. Public disclosure of our profitability would have resulted in even more intense price reduction pressure than normal, and "normal" was already extremely intense. The idea of doing an IPO was never seriously considered again.

At the June 30 conclusion of fiscal year 1981, sales were $12,117,000, up 29%. Net after-tax profit was $844,000, up 16%. Long-term debt jumped to $2,800,000; but, considering the prospects and a book net worth of almost $5 million, it was not excessive. As anticipated, the semiconductor LED business declined to $807,000, less than half of what it had been the previous year. Calculator makers were moving to liquid crystal displays. Fortunately, components and systems volume doubled to $1,209,000. The Honeywell auto-focus business passed its peak and declined 33% to $700,000. Massive-optics PTV sales jumped $4,230,000 to $9 million, up 89%.

Calendar year 1981 continued to be a difficult time economically. President Reagan, his new team and the Federal Reserve were doing everything they could to beat down the inflation rate which was 10.3%. The bank prime lending rate reached 20.5%, and for the year averaged 18.9%. The dollar continued to be very strong against the yen, reaching a high of 261. It was also very volatile, with a 24% difference from the high to the low. That meant the price of our lenses for Japanese customers varied 24%, even though the price in dollars never varied. More and more, we worried about the risks to USPL inherent in the dollar-yen currency exchange relationship.

As we were about to enter calendar 1982, the last half of 1981 (which was the first half of fiscal year 1982) had been very strong. USPL had 600 employees with over 100 still on probation because they had been there such a short time. Customers were excited about Compact Delta I, but the thermal stability weakness remained. It was a fundamental problem that needed to be solved, and we did not yet have a good answer.

# 44 Vivitar

In the 1960s, Japanese photographic equipment makers were developing impressive camera designs and manufacturing capability. United States makers were rapidly losing their limited market position. German cameras were generally regarded as having the highest quality, but their makers were also beginning to suffer from Japanese competition. Companies such as Canon, Konica, Mamiya, Minolta, Nikon and Olympus were challenging the great German names of Exakta, Leica, Rollei, Voightlander and Zeiss, as well as Sweden's Hasselblad, for market share.

The Japanese were producing excellent lenses, coatings and camera bodies at very low prices, although probably not the best. The price-performance equation was outstanding. Their problem in the big United States market, was marketing, selling and distribution. They did not have the skills to perform these functions for themselves, so they relied on American distributors to do the job.

Ponder & Best, a Los Angeles photographic products distributor, sold Rollei cameras to photo equipment stores in the western United States. A similar company handled the eastern section of the country. Once Rollei's market position had become established, it set up its own distribution function and eliminated Ponder & Best, who in turn countered by taking on the Japanese Mamiya line with national coverage. Mamiya, like Rollei before them, took the distribution in-house once their position had been established. It became clear to Ponder & Best that, eventually, other major brands they handled would do the same thing. They needed a better strategy if a sound long-term business was to be built.

The creative thinking to find a solution came from Bruce Shomler, the executive vice-president. He decided Ponder & Best needed to buy products, plus develop and have products made to their specifications, that would be sold under their own brand. The brand name they chose was Vivitar. Not long thereafter, they also changed the name of the company to Vivitar.

In a risky move that became strategically important in time, Shomler bought circuitry patents for a flash system from an obscure Belgian inventor. Eventually, all Japanese flash makers would license the technology. Another significant brand addition was the Olympus line of cameras and lenses.

Shomler somehow learned that Ellis Betensky was a creative lens designer, and he met with him in 1968. The following year Ellis started Opcon Associates, and Vivitar quickly became his most important customer. Unique patentable designs were created by Opcon and built by lesser known Japanese lensmakers which, with aggressive promotion, allowed Vivitar to gradually build a reputation with photo dealers. Their zoom lenses were particularly recognized to have excellent performance and features. Photographic magazines conducted comparative tests and usually reported glowing results on Ellis's designs. The lenses could be used on any major Japanese maker's 35mm camera.

Ellis was intrigued by the possibility of incorporating plastic aspheric elements into photographic zooms that frequently contained up to ten or twelve elements in their all-glass designs. He suspected performance could be increased and costs simultaneously lowered through element reduction. He had been telling me about Vivitar and this possibility since about 1975.

In 1976, he encouraged me to join him in attending the world's largest photo products show—Photokina, in Cologne, Germany. I agreed. We stayed with a large group of senior Vivitar people, which gave me a chance to meet the key executives. Ellis took me through the gigantic exhibition halls for two days, providing a tutorial on all the players and their relative position in the world of photography. This began a very active relationship with Vivitar that lasted nearly ten years.

Ellis's first idea on how plastic aspheres could be incorporated involved the use of one element in a design. The Japanese lens supplier that would be required to assemble it vehemently objected to USPL's involvement in the project, and after much negotiation the idea was dropped. I later suspected that Vivitar received concessions as part of the agreement not to proceed with our lens.

In anticipation of other Vivitar projects soon thereafter, we purchased a high quality, specialized European molding machine with many process control features for the purpose of doing photographic lens research and development work. Much effort on our part went into building a photographic lens molding capability.

Vivitar was importing low-end, small-format 110 cameras and selling them in large numbers under their brand name. They became the dominant supplier of a line of flash attachments that were considered superior to the competition, and Vivitar's interchangeable 35mm lens products continued to grow in reputation as Opcon kept delivering new innovative designs. Vivitar was so successful they decided to build cameras and lenses in California at a yet-to-be-built facility that would be called Vivitar Park. It was somebody's heady idea to emulate Eastman Kodak's name for its mammoth manufacturing complex in Rochester called Kodak Park. For a company that had never made anything, it was quite an ambitious plan.

A team was hired to get them into the manufacturing business. The new group leader requested that I visit Los Angeles to discuss our involvement. Within minutes of meeting him, he proposed that we prototype the lenses, show them how it was done, and help them get their own lens plant running. He said it was a chance for us to make some money on something they were going to do for themselves anyway. At this time, after investing over two years in Vivitar dedicated development equipment and trials, we had no business to show for our effort. The talk was always about our great opportunity in the future. Now, as if with a flick of his hand, this fellow was saying his proposition was a good deal for us because they would do it with or without us. I was furious, and told him we had no interest in participating. I also told the Vivitar executives with whom we had spent so much time and who had courted us, how disappointed we were in their manufacturing leader's behavior. They were embarrassed, but they had also committed themselves to a course of building manufacturing capability. I thought that was the end of our relationship.

Only a few months later, Vivitar's business turned down and Olympus set up its own distribution arm. They were under great financial pressure, which led to a major restructuring. That was the end of Vivitar Park and the team that was going to make it happen.

They had yet another problem. The Japanese lensmakers that had been building Opcon designs were with minimal changes, now making lenses under their own names and competing with Vivitar as well as supplying it. To gain leverage by having a non-Japanese source, in about 1980 they signed a long-term contract with a Taiwan lens-making subsidiary of the fine German company—Robert Bosch. The subsidiary was called Bauer & Sun, and had been acquired by Bosch when they bought the German Super 8mm movie camera maker, Bauer. The Super 8 business was dying with the rise of video cameras, so the deal with Vivitar was an answer for the Bosch problem of finding business for their factory that everyone referred to as BASO.

In 1981, after apologizing for the Vivitar Park fiasco, Bruce Shomler convinced us to develop a large aspheric plastic element that was key to building a revolutionary new Opcon design for a telephoto lens. Because of our past experiences, I was doubtful about the wisdom of participating. However, the extraordinary features of the new lens made the idea compelling. It was a catadioptric (referring to a lens utilizing an internal magnifying mirror) 450mm design that was f/4.5, only 5.7 inches long, and 40% lighter than the much larger f/8 all-glass telephoto competition. (For those unfamiliar with photographic lenses, f/4.5 was very fast for a lens with such a long focal length. The speed allowed significantly more light to reach the film surface. This in turn allowed exposure time to be reduced by using a faster shutter speed. This minimized the time needed to hold the camera steady.) The 450mm focal length meant that objects far away could be magnified to appear extremely close. This combination of magnification, speed and light weight represented a design break-through that couldn't be achieved without a plastic asphere made by our massive-optic technology. It was code named MUM. Photo-graphic magazines and *The Wall Street Journal* hailed it as an important development, giving credit to Opcon and USPL.

All parts, except our plastic element, would be made at BASO in Taiwan. We were urged to visit BASO, so in March of 1982, after calling on customers in Japan, and before a trip to China, Brian Welham and I went to Taiwan to meet with them and see the factory. Vivitar didn't know it, but we had an ulterior motive. The growing thermal stability problem of our Delta lenses had not been solved. We were searching for United States sources of glass element supply and suspected BASO might be an Asian possibility.

The BASO factory was in Taichung, Taiwan's third largest city 100 miles south of Taipei. The Bosch people that managed it had created a clean efficient facility. The number two ranking manager was an impressive Austrian physicist, Dr. Peter Scherb who clearly was the most important person to the operation. After an extensive factory tour, meetings were held to coordinate our respective activities.

At this point in time, after six years we still had not done any business with Vivitar, and I still had my doubts about the wisdom of continuing to work with them. Nevertheless, the BASO visit would lead to an agreement in eighteen months that would be critical to our future success. There was much more in the Vivitar story still to come.

# 45 China

Brian and I departed Taiwan for a long weekend in Hong Kong before traveling to China. We spent an evening with one of our Beta lens customers and otherwise toured one of the most vibrant cities in the world. It was my third visit, and my fascination for this unique entrepreneurial place never diminished.

We stayed at the spectacular Peninsula Hotel, which many people believed at the time was the finest in the world. A stay at the Peninsula began when they picked you up at the airport in a Rolls Royce. The service was incomparable. On the bed headboard were a series of buttons to press for a variety of services such as housekeeping, valet and room service, plus many more. After a few hours sleep the first night, when rolling over my arm inadvertently swung back against the buttons, and soon people were coming from everywhere to see what I needed. In my dazed condition, I couldn't figure out what was going on until the corridor houseboy stationed nearby explained what undoubtedly had happened. It was quite a gathering in the middle of the night.

Some months earlier we had been contacted by William Wright, a principal in East-West Services, a Washington and Shanghai-based firm that was created to export U.S.-made technical products into China. We were not enthusiastic about using sales agents, but realized that if business was to be done in that country, one would be necessary because of all the government bureaucracy requiring complex specialized procedures.

Bill Wright visited USPL and convinced us we had nothing to lose by letting him represent us in China. Soon thereafter he reported a projection TV lens opportunity in Shanghai and encouraged us to visit. This is why Brian and I were making the trip.

The Monday morning flight was non-stop from Hong Kong to Shanghai on an old Russian-built jet in such poor condition that we had serious trepidations about what we were doing. In spite of its oily exterior surfaces, broken seats and loose interior parts, we nervously went on with the journey.

We were met at the airport by Bill Wright and a delegation of five people from Radio Factory No. 32—the state-owned 1,500 employee facility that was planning to make projection TVs. In the delegation was Madam Tu—the thirty-four-year-old engineer who was the plant manager, the chief engineer, the Communist official who oversaw all factory activities, and a couple of other people. In rough-running old Russian-built cars they took us to the Jin Jiang Hotel for check-in, where we had lunch by ourselves before the afternoon meetings. This was the hotel where President Nixon, on his ground-breaking trip, and Chinese Premier Chou En-Lai issued the Shanghai Communique in February of 1972 that delineated all the issues upon which the two countries could agree.

In 1982 there were no new hotels in China. Shanghai was the third largest city in the world with twelve million people, and showed no signs of recent factory, office or housing construction. Except for a select few, virtually everyone traveled on bicycles. The Jin Jiang, having been built by the French earlier in the century, was very attractive architecturally. Unfortunately, all the furnishings appeared to be pre-World War II. Nevertheless, it was quite acceptable as our home for the next few days.

After lunch we went to a large meeting room where approximately fifteen Chinese engineers waited to see us. Bill Wright introduced us, giving background on USPL and our position in the world TV market. Shockingly, he then told them we would conduct a seminar on building projection TVs. This was the first time we had heard about it, and with no preparation it was a very unsettling situation. All we had to work with was an old-fashioned blackboard and chalk. The Chinese engineers' English skills varied from none to fairly good in a couple of cases. Since we knew no Chinese, it was clear the afternoon was going to be interesting and painfully long. I talked about USPL and how we built our television business. Then Brian took over the bulk of the presentation, outlining to them all the things that needed to be done to build a projection TV with emphasis on optics and CRTs. Under the conditions he did a spectacular job of narrative, combined with outlines, drawings and mathematical formulas on the blackboard. Every Chinese engineer dutifully took many pages of notes during this first meeting.

That evening we were entertained with what Americans would call a very nice dinner that the Chinese termed a banquet. In addition to Madam Tu and the Communist official who watched and kept tabs on every official move she and others in the factory made, we were accompanied by senior engineer, Bill Wright, and the government bureau chief of all communication activities in the Shanghai region. The table was round, and in front of each place there were three glasses. A large one contained beer, a medium-sized one was filled with plum wine, and a small liqueur-type was filled with a clear, nearly 100% alcohol called Moutai. The bureau chief was a very important man in Shanghai, and he took charge of the dinner. As he smoked literally non-stop, he encouraged us to drink plentiful amounts of what was before us—especially the Moutai. I remembered reports on President Nixon's 1972 trip where he could barely drink the Moutai because it was so strong. Sitting there, I thought this stuff would have been an illegal fire hazard if it had been found in one of our factories. Over twenty courses of food were brought to the table. Adventurously, Brian and I had a little of everything, not inquiring about what it all was and not really wanting to know.

The bureau chief had such a good time that he invited us for another banquet the next night. The thought of more Moutai was too much for us. Goods not available to Chinese citizens were sold to foreign visitors in what they termed Friendship Stores. There was one in the Jin Jiang complex so I went to it to find a bottle of Kentucky bourbon as our contribution to the evening, hoping the fact that it came from near Cincinnati might make it not only a nice thing to do but also an acceptable substitute for the Moutai. Unfortunately, the only thing I could find was a blended whiskey called Four Roses. I had never tasted it, but remembered from my youth hearing that it was low cost and quite popular. At the banquet that evening everyone was delighted with the gesture, and, together, we all tasted Four Roses for the first time. It was unbelievably awful—worse than the Moutai! Now I had a problem. These people, not wishing to offend us, were going to be forced to drink something that clearly none of us could stand. To everyone's relief, I immediately declared it a huge mistake, and toasted our host with the Moutai which, in itself, ensured that Brian's and my nasal passages would never be the same. After the banquet we were taken to see a performance by incredibly talented Chinese acrobats.

We were told by the communications bureau chief, that the Chinese government had concluded that projection television might be a good way to communicate with the population that then totaled well over one billion. The thought was to put up satellites that would broadcast to the hundreds of thousands of communes that then existed. There were so many people in each commune that viewing on individual TVs was out of the question, and normal CRT television sets were too small for large groups to see. The statistics made the idea of projection compelling.

Radio Factory No. 32 had been chosen as the facility to develop PTV. When Brian and I visited, we were instructed to take off our shoes and put on open toed slippers. Then, in what seemed like a contradiction, we walked through dirt floored corridors to get to the part of the factory where 100,000 small black and white TVs were being assembled annually. In a development laboratory nearby, they surprised us by showing Schmidt-type black and white projection CRTs that had been experimented with earlier. Now they wanted to upgrade to color and had made a deal with Sony to buy components which would be assembled into sets similar to the Sony coffee table model which used Delta II-Ms. With this development, the idea of selling the Chinese became more attractive because it was just a matter of assembly and the purchase of one of our standard lenses for them to be in the business.

Radio Factory No. 32's problem was getting currency released by the government to buy the lenses. This required our dealing with yet another bureaucracy. Together with Bill Wright and one of the representatives of No. 32, we spent nearly a day trying to get a letter of credit issued for 100 kits consisting of three Delta II-Ms and fifty Beta lenses that they wanted to use for the development of single-tube systems. Eventually we obtained verbal approval and were told the paperwork would come to us when we returned to the United States.

Finally, our five-day visit was about to come to an end. With travel to Japan, Taiwan, Hong Kong and China, we had been gone eighteen days. For the last few, I had been promising Brian that the Pan American Boeing 747 that was going to fly us home would have a nice standing rib roast on board that would be cut to his satisfaction. That kept him going! The day before departure we had a final meeting with the folks from Radio Factory No. 32 in our suite of rooms at the Jin Jiang. I bought a box of chocolates at the Friendship Store that were to be served at the meeting. Shortly before the session, Bill Wright informed me that it was absolutely forbidden for the Chinese to take any personal gifts and the chocolates would be a problem. With this in mind I waited until near the end of the session and then opened the box, saying that passing out candy like this was a custom of ours. When after a couple of minutes no one made a move, I said, "I would like you all to take some for your children." After a long pause, the Communist overseer took a few pieces. This made it all right for the others to take some, and they eagerly did. I have always wondered if the Communist's chocolates were actually delivered to his children.

We were invited to Radio Factory No. 32 for a typical employee lunch just before our departure. It was a very cold March day, and with no heat in the factory we kept our overcoats and gloves on. The lunch was a hot noodle soup with a little bit of chopped eel sprinkled on top for flavor. Bound in heavy clothing and under frigid conditions, we tried to eat the noodles with slippery chopsticks. It was a messy disaster. That was the end of two good neckties and represented yet another interesting experience.

We left with hope for China becoming an important television lens market, but also much doubt about their ability to part with the necessary currency considering all the demands they had for it. In November, Dave Hinchman spent a week in Peking—the capital now called Beijing—working with Bill Wright calling on the companies in that region. Not surprising to us, after the first order we never did any more PTV lens business in China. However, the visits did provide us with valuable education that was useful in the future for evaluating opportunities that were presented to us by various people to do business there.

# 46 1982

By the beginning of the year, Vivitar was convinced plastic aspheric lens elements combined with glass elements, would provide performance and cost reduction opportunities that would lead to photographic lens competitive advantage. They were the largest sellers of interchangeable single lens reflex 35mm camera optics. We had signed a five-year "program of cooperative development" that they hoped would lead to purchases of $10 million to $20 million from us within five or six years. Based on our earlier experiences, we continued to have doubts that Japanese photographic lensmakers would incorporate USPL aspherics into Vivitar hybrid systems designed by Opcon. There were other reasons for doubt, including Vivitar's ability to coordinate such a development. The agreement called for us to sell to Vivitar on an exclusive basis. This was acceptable because Polaroid and Kodak had their own aspheric capability, and the only other potential customers were camera makers in Japan who clearly did not want to buy photographic aspheric elements from us. At the time, our only Vivitar activity was the top priority 450mm f/4.5 telephoto MUM development that was to be built at BASO in Taiwan.

Meanwhile, Sony surprised the photographic world by announcing and demonstrating a non-film-based digital camera that electronically recorded and stored photos. They could be viewed on television sets or printed by the high quality computer printers of the day. Sony named it the Mavica. The picture quality that could be attained with this pioneering product, frankly, was not very good. However, all knowledgeable observers knew it was a technology that would eventually be perfected to a level where it would compete with traditional film photography.

During a March visit to Sony, senior managing director M. Morita, who was in charge of their worldwide video business, asked to visit with me. He said that Sony had high hopes for Mavica-type products and a new generation of small format home video cameras. Then he handed me prototypes of different models they had made to emphasize his optimism. I had heard of the Mavica but not the new video camera ideas. Mr. Morita further explained that Sony would like to explore a closer relationship with USPL. His proposal was for us to make aspheres that would be incorporated in hybrid lenses assembled in Japan by a lensmaker in which they had a financial interest. He asked for an exclusive arrangement that called for us to produce such lenses for Sony only. This, of course, was a problem because we already had an exclusive agreement with Vivitar. Sony was one of our largest and most prestigious customers, and unless we could obtain a release from Vivitar we had to decline its request. Reluctantly with as little detail as possible, I told him about the Vivitar commitment and our need to get a contract modification. We left on the basis that I would talk with Vivitar and further coordinate activity relating to his proposal with Sony New York-based senior executive, Michael P. Schulhof.

There were many things about this that concerned me. We had not yet demonstrated that we could make photographic aspheres at the precision levels required with conventional molding techniques. If we struggled with this project, it could lead to a replay of the compact disk lens effort. The risk of anything less than total success would reflect negatively on the great relationship that existed as a result of our far more important projection TV lens business with Sony. All these concerns were forthrightly explained to Mr. Morita prior to any discussions with Schulhof. Michael "Mickey" Schulhof was an American highly trusted by the Tokyo management. He was not so highly trusted by some of the American Sony executives. When one of them learned we would be attempting to negotiate an agreement with him, I was called and warned with the words, "Be very careful in anything you say and agree to with Mickey." Never before or since have I received such an admonition from a customer about someone in his own company.

Vivitar agreed to release us from the contract, although narrowly, to accommodate Sony. During arduous negotiations, we concluded an agreement to do essentially what Mr. Morita had initially proposed. I still had doubts about the wisdom of our being involved, and it would seem they might have also had some concerns because very little follow-up activity ever took place.

Mickey Schulhof's high-profile career continued to blossom at Sony. He was instrumental in acquiring the CBS record business and soon thereafter became the only American appointed to the Sony board of directors. As the top Sony executive in the United States, he led the company to buy a major Hollywood motion picture company. Unfortunately, under his leadership, most of the deals turned out to be disasters. Years later, when Sony's Tokyo top management changed, one of the first acts of the new chief executive was to fire Mickey. That was followed by a colossal multibillion dollar writedown of the entertainment assets Mickey had acquired. It was one of the biggest writeoffs ever to that point in time.

The 1981 tax act liberalized rules for corporations with a small number of shareholders allowing them to be taxed like a partnership. Most corporations pay taxes on earnings and then, when dividends are paid, the recipient is taxed again at his or her regular rate. This double taxation is onerous and a significant factor in causing companies to declare small dividends.

The new tax rules made it possible for us to become what is called a Sub Chapter S corporation. This meant that, like a partnership, all earnings flowed through the company without tax proportionally to shareholders who then paid one tax at their individual rates. Sub Chapter S corporations, in addition to providing limited liability to the owners, could retain earnings for corporate use which, when paid as dividends, were subjected to no further tax. There were other advantages including the fact that the maximum individual tax rate was lower than the corporate rate. Every feature of Sub Chapter S status was better for us. We converted and in the future would pay dividends in an amount to cover shareholder tax liability.

The first half of fiscal 1982 (July 1, 1981 to December 31, 1981) was a phenomenally strong period for USPL. Unfortunately, by January we were experiencing precipitous drops in television lens orders as the manufacturers' Christmas sales were disappointing, leaving them with large inventories to work off. Dave Hinchman reported to our associates in the May issue of *FOCUS*, that since January, our shipping level had been less than half of what was experienced in a similar time frame in the fall. Virtually all customers had been designing and engineering more desirable, smaller cabinet sets made possible by the new Compact Delta I. After the slow fall sales to their retailers that had just been experienced, they did not want to have an excess inventory problem with older goods when the new compact models were introduced in September. To accomplish this they drastically cut prices—frequently to below cost. We understood why they had to do it but, were nevertheless, worried that consumers were receiving false signals as to what PTV would sell for in the future.

The great beginning of fiscal year 1982 provided the boost that carried us to $21,856,000 in sales—an increase of 80% over fiscal 1981's $12,116,000. Net after-tax profit was $2,170,000—up 157%. Projection TV lens volume over the previous year was up 113%, or $10,182,000, for a total of $19,188,000. Semiconductor and Honeywell auto-focus systems were down somewhat, while components and systems held steady over the previous year. Because the standard optics manufacturing people were consumed in supporting massive-optics with the production of lens housings, we were not concerned about flatness in that part of the business.

With such a "blowout" record fiscal year, we should have been a lot happier than we were when it finally came to an end. The tough second half, accompanied by economic conditions that were still quite nasty, took most of the joy out of it. The bank prime interest lending rate in 1982 continued to be near historically high levels, ranging from 11.5% to 16.6%. The yen to dollar exchange rate bounced between a low of 219 to a high of 278—a difference of 27%. At 278, it was the highest we had experienced since doing business in Japan.

In August, Dave reported on aggressive competition we were experiencing in Japan from lensmakers using historically more expensive conventional glass technology. This was made possible because of the grossly undervalued yen, causing us to reduce prices on a number of the Delta products to defend market position. To bring costs into line with the reality of the business level, spending was cut everywhere possible by mid-year. With deep regret, employment was reduced from over 600 to slightly less than 400. The unanswered questions causing us worry were how much the high-interest-rate economy would affect buying patterns and how new model Compact Delta I projectors would be received. The only favorable economic news was that inflation, measured by the consumer price index, was dropping from 10.4% in 1981 to 6.1%—still an historically elevated level.

It was a bewildering year for us and the projection television manufacturers. In calendar 1981, PTV sales to retailers had more than doubled. As 1982 began, USPL, *Television Digest*, and the Electronic Industries Association (EIA) were forecasting another year of over 100% growth. When the final December numbers were reported, the growth had only been 36%.

We and others speculated as to why consumers were so cautious. With the unemployment rate jumping from an already high 8.2% to an alarming 10.7%, there was a reluctance to purchase "big ticket" durable goods. Also the 1981 tax act encouraged personal savings with the creation of Individual Retirement Accounts, All Savers Certificates and a tax-free utility dividend investment plan. These drew enormous amounts of money into savings that might have been spent on luxury goods such as PTV. Statistics show the personal savings rate in America, on a quarterly basis, fluctuated near historically high levels of between 8% and 10.2%.

Fortunately, fiscal 1982 had been very profitable because $2.5 million for capital expenditures had been spent and the jump in sales required over $1.4 million in additional working capital. When the sales softness was recognized, the two places we did not reduce spending were product and technology improvement. More sophisticated lens designs, as a result of customer demand for higher performance, required the massive-optics process to be continually improved. Improvement meant optical surfaces formed to higher tolerances with less deviation from what was theoretically perfect design. In April, Brian Welham reported to the board on a new $100,000 machine internally developed by our engineers. It was capable of producing optics within astonishingly high tolerance levels of ten one-millionths of an inch. Small upgrades were being made in virtually all processes that added up to significantly higher quality. The push for continual improvement was a phenomenon that was always with us. Now it was intensifying.

At the Chicago 1982 Summer Consumer Electronics Show, Magnavox was the first company to introduce a projection TV with Compact Delta I lenses. It was a 37-inch set that gave the customer twice the viewing area of the largest conventional CRT TV which was 25-inches. Not all TV manufacturers had been informed about the development of Compact Delta I because it was clear some preferred to purchase from Japanese glass lens sources. The Magnavox offering had a very attractive cabinet that was extremely small, with less depth than 25-inch CRT models. It was great fun for us and the Magnavox product engineers to watch engineers from competitive manufacturers, who didn't know about our new space minimizing development inspect the models at the show. A number of them made tape measurements on the cabinet and then just stood there shaking their heads, bewildered with what they were viewing. The Magnavox product introduction, utilizing Compact Delta I, was considerably ahead of competing manufacturers whose offerings came to market a few months later. Most product design people in the industry thought a screen size of 37 inches was too small and that the minimum should be 40 inches. In time, that view proved to be correct. The Magnavox set was moderately successful, but all future models by it and others would be larger—with the most popular size being 45 inches.

The impressive USPL TV lens production ramp, in the first half of fiscal 1982, could not have been accomplished without an enormous effort by the standard optics division in supplying housing tooling and housings into which the lens elements were assembled. In that year it produced $234,000 worth of complex molds and $2,785,000 of housings used internally with transfer pricing at cost. Our people had the capability to rapidly design, make changes and build the high quality and large volume of parts that were critical to successfully producing the final product. This internal resource gave us further competitive advantage and would always continue to do so.

Throughout the year we worked with our old friend, Milt Leonard, on a revision of *The Handbook of Plastic Optics*. A new color brochure was produced, and the first significant effort to generate new standard optics business with a series of trade magazine advertisements was attempted. Although over 2,000 brochures were sent out to ad respondents, very little direct business could be traced to the effort. It reinforced, in my mind, the importance of our earlier successful technique of placing plastic optic technical stories in top trade magazines that carried our name as well as Cincinnati location, thereby encouraging potential customers to contact us.

As Compact Delta I went into production, Brian and his engineering team were doing development work on the next generation of lenses. We were studying the optical coupling of the projection lens to the CRT to eliminate the air gap that caused reflections which further caused a loss in picture contrast. It was an intriguing idea because cooling of the CRT with a liquid at a compatible refractive index not only dramatically improved contrast by eliminating the reflections but also theoretically could reduce the serious continuing problem of focus shift caused by the expansion of power elements as the CRTs heated the lenses during operation. RCA engineers and others came to see our earliest demonstrations. Most of the TV manufacturers began research on the idea.

In addition to Philips, Thomson and Barco, we added Grundig to the list of European projection lens customers requiring visits to its headquarters in Nuremberg, Germany.

By November, it had been over a year since we agreed to build the 63,000 square-foot Kentucky factory, after the state constructed an entrance and exit interchange off of Interstate 275 at Mineola Pike near the Greater Cincinnati-Northern Kentucky Airport. Kentucky officials had not contacted us with plans or build schedules. With the severe business downturn, customers' concern about our one location vanished. The new priority was how purchase order commitments could be reduced. We had an agreement to honor in Kentucky but, given the business climate, there was no immediate need for the facility.

Finally, near the end of the month, a Department of Highways official sent detailed drawings of a plan. We were amazed to see that it required the state to acquire nearly half of our land, making the site unusable given the building design and parking requirements. They exhibited no flexibility and proceeded to buy from us what was needed. We put the remaining portion on the market and sold it a few months later to a Holiday Inn developer, with the net result being a profit in excess of 100%. It was ironic that after pushing so hard to get us to build at that location, they then made it impossible to do so. That was the end of thoughts about building in Kentucky.

To provide for future expansion, we acquired ten acres directly across McMann Road from the headquarters location. This parcel would be added to a couple of times in the years to come, and eventually totalled fifty-nine acres.

I had not seen Henry Buhlmann, the man who had the idea for plastic optics at USPL in the 1950s, since 1971. I had thought about him often because if he had not done what he did none of us would have been doing what we were doing. On the premise that he might enjoy seeing what had happened to his old company, I called to see if he would like to come out for a visit. He eagerly accepted, and on August 17 was brought by a friend. At eighty-three years old, Henry was very sharp in mind and extremely interested in where we had taken the enterprise. The company was now well over 100 times the size it was when he had last seen it. As we toured through both factories, it was fun to introduce him to our people as the man who planted the seed that was now enjoyed by so many. After the tour we had our pictures taken together, then had a long chat that ended with him telling me how delighted he was because he had always hoped someone would take his early creation and build it into a significant business.

This visit was the last time I saw Henry Buhlmann. Sixteen years later, he died a few months short of his 100th birthday on August 5, 1998. His passing gave me the opportunity to again credit him in his obituary as the person who had the original idea that eventually benefited so many people in our community and industry.

# 47 Projection Television Past and Future

By the beginning of 1983, there were many consumer electronic magazine writers and industry executives commenting on the future of projection television. In addition, research organizations were doing surveys which led to supposedly authoritative reports that were being offered at high prices. Most of what we read on the subject was highly flawed. This was a concern because we believed it was important for the TV manufacturing executives to be making decisions based on solid information. Our fear was they would become disillusioned with PTV if market failures resulted because actions had been based on erroneous data.

When dealing with the manufacturers we were careful to accurately give our views on how the industry was developing without, of course, revealing any one company's proprietary information. Because we were selling lenses to the majority of the PTV makers, and frequently calling on those who used competitive lenses, our vantage point as an observer was unique in the industry. No other company was actively involved with all the players.

In 1982 General Electric installed Jacques Robinson as the general manager of its television business, with the admonition from the CEO to make it a good profit-maker or they would get rid of it. Robinson came from another part of the business, so we didn't know him. By the end of the year he was giving interviews to the consumer electronics industry press, which usually included an industry analysis of what was happening with PTV and predictions for its future. His comments were not even close in accuracy to what was actually happening, and the predictions were wild. Customers began calling us to inquire as to why the much-admired General Electric Company had such a divergent view of the business.

By December I was becoming seriously perplexed by the things Robinson was saying, and to a lesser extent by the comments of a few writers. With disappointing year-end PTV sales, and all the inaccurate rhetoric in the air, many important people in the industry were confused and beginning to have doubts about what to believe. I felt our interest in the industry information being credible required that we take a proactive role in providing the data and analysis.

During the latter part of December I wrote an eight-page paper entitled "Projection Television Past and Future." It began with an introduction as to why we thought it was needed and an explanation that it was prepared with numerous inputs from knowledgeable people throughout the industry. It contained sections on:

History Of Projection Television

Growth Of Projection TV Market

Consumer Sales And Dealer Inventories

1982—A Year Of Disappointment

Why Wasn't 1982 A Bigger Year?

Confusing The Consumer In 1982

1983 USPL Forecast

How Projection Has Changed

Why Projection Will Get Much Better

Quality Projection Sets In The Market

Belief In Projection By Manufacturers

1984—A Year With Positive Factors

Dave Hinchman and others reviewed the copy, made editing suggestions and helped me compile the data.

At the beginning of January I called our friend Dave Lachenbruch, the highly respected editor of the industry's most authoritative publication, *Television Digest*, to tell him about the paper and why we thought its publication was necessary. He agreed that it was needed. He then asked if we would send an advance copy that *Television Digest* could review and summarize in its next weekly issue. We were delighted because *Television Digest*'s endorsement would give the effort additional credibility and inform industry people all over the world of the paper's availability.

We took a large quantity of copies to the January Las Vegas Consumer Electronics Show and gave them to key customer contacts along with an explanation as to why we had been moved to write it. Simultaneously *Television Digest* did a highly favorable review, and soon requests from many people we never had heard of started coming in. The reaction was universally positive. We had accomplished my objective of providing a sensible analysis of what was transpiring in the world of projection television. From that point on industry reporting was more factual and forecasts were more rational.

I had envisioned the paper as a one-time publication to correct a specific problem of the moment. However, as the year went on, quite a number of people asked when the next edition of "Projection Television Past and Future" would be published. Japanese customers were particularly desirous for us to update it. Sometimes during sales calls people would bring it out and use it as a reference point for discussion of industry trends.

Because so much encouragement to produce revisions was received, we decided to make it an annual publication. As this is being written, the seventeenth edition has recently been issued. It has been continued in a format similar to the original with appropriate additions. Now well over 1,000 copies are distributed annually in both the English and Japanese languages to key people at television manufacturers and PTV component suppliers throughout the world.

The publication has continuously enjoyed a reputation for being authoritative; plus, it has enhanced the industry's view of USPL. If the General Electric head of TV manufacturing had been a little more accurate on the industry facts in his interviews and not quite so wild with his forecasts, there would have been no "Projection Television Past and Future."

# 48 How We Viewed USPL As a Business

By 1980 it had become clear that projection television would develop into an important consumer product. Our invention of the massive-optic aspheric lens manufacturing process, combined with the unprecedented Opcon Delta lens designs that could be made with it, added up to a significant advancement and competitive advantage for the company in PTV optics.

Conceptually, for a number of years we viewed USPL as a plastic optic technology company that made most of its profit by supplying unique or near unique components principally used in advanced consumer products. The examples were Polaroid's SX-70 camera, hand-held LED calculators, digital watches and auto-focus cameras. The design, manufacturing and precise measuring requirements for multi-element aspheric lenses caused us to ratchet up the technology designation to "high technology." No company had ever mass produced aspheric optics resembling ours for PTV or any other application.

Given the life cycles of the markets and the tumultuous economy of the 1980s, we also viewed the business as being potentially very volatile. Fiscal 1983 (July 1, 1982 to June 30, 1983) was a severe reminder as to the correctness of that observation. This all led to a well understood policy that financially, we would balance the volatile fast-changing consumer markets against a very strong financial position. The realization that USPL was a high risk technology business caused us to frequently reiterate that only growth opportunities that were a logical employment of the company's skills would be pursued. We would not try to grow for growth's sake alone—only if something special was brought to the equation. The dangers of getting involved in activities foreign to the company's specialty also were well understood.

The simple theory of only pursuing growth that related to our skills, accompanied with a strong balance sheet, meant that in bad times we would have a financial cushion to fall back upon, and when really good opportunities came along the company would have the financial strength to aggressively move on them. There was no better example than the sound balance sheet in the late 1970s that was, for the most part, built during the LED magnifier boom. It allowed us to aggressively develop the TV lens technology. Another was the nearly $4 million increase in capital expenditures and working capital that was required to support fiscal 1982's 80% sales increase from readily available cash.

At the beginning of 1983, with a net book worth of $8 million and $2.5 million in long term debt, USPL had $6.3 million in cash. The large cash position allowed us to comfortably look beyond the current market turmoil and do what was necessary to prepare for better days that could be foreseen.

Occasionally a few friends who knew our numbers would criticize me for having too much cash. I found comfort in a strong balance sheet. Frankly, I never thought having a lot of cash was a problem.

# 49

## Thermal Stability:
## It Had Become a Serious Problem

Customers were driving PTV CRTs harder to obtain higher picture brightness. This generated more heat, which in turn further exacerbated the problem of the all-plastic Delta lens power element expansion, which caused a slight defocusing of the system. To compensate, PTV makers ran the sets in the factory until maximum temperature was achieved and then focus the heated lenses. This meant that when a buyer turned on his or her cool set there was slight defocus until it had been run awhile.

The TV makers were growing more impatient with the flaw, and the pressure on USPL to fix the problem was increasing. Our efforts to solve it with design changes produced only minimal results. To make matters worse, Japanese lensmakers were becoming aware of the limitation and beginning to claim the stable focus of all-glass systems as a significant advantage. They were correct. In mid-1983 Mr. Yoshikawa, the manager in charge of television for Nippon Electric Corporation (NEC), made it very clear that if we didn't correct the problem in a few months his company would obtain its lenses elsewhere.

Our 1982 effort to find glass element suppliers in the United States was fruitless. Companies that made military lenses and overhead projector lenses quoted prices that were totally unacceptable. Brian Welham had obtained quotes from Bausch & Lomb that were extremely high. Good sources were not available because the lenses required were quite thick and large in diameter. There was very limited manufacturing capacity for such designs in the world except at the Japanese PTV lens competitors. By March of 1983, it was concluded that we would probably have to develop internal capability and capacity.

The prospect of quickly becoming a glass lens maker was daunting. Although optical fabrication had been an area of concentration for Jim Bohache at the University of Rochester, he did not have high volume manufacturing experience. No one else in the company did either. Another problem was finding domestic sources of large glass blanks—the raw material of a lens grinding, lapping and polishing operation. The final concern was that glass lenses would require coatings that were applied with a slightly different process than we were using on plastic. Nevertheless, there was no option other than to plunge into a study of how we could get into the business.

Our German customer, glass lensmaker Schneider Optical in Bad Kreuznach, had fallen on hard times and was going through a reorganization. Raimund Muhlschlag, an engineer who worked closely with us, wrote with the news that he was out of a job. This was a surprise, as he had been with the firm for many years. I called to discuss his situation and learned about German glass lensmaking equipment manufacturers. We were particularly interested in DAMA and LOH, both of which had excellent reputations. After I told him of the large diameters we needed to process, he offered to contact the two German companies to obtain recommendations, costs and capacity throughput data.

Brian and I visited Bausch & Lomb to determine if it could be a source of lens blanks. This trip was not encouraging. In April Jim Bohache and I visited the Corning Glass Works, in Corning New York. The people we met with were somewhat encouraging, but they never followed-up with detailed information. The only other possible U.S.-based supplier was Schott Glass in Duryea, Pennsylvania.

We planned a visit to Schott and a highly-regarded glass lensmaker also in Pennsylvania—Plummer Precision Optics. Ray Muhlschlag had collected a lot of information for us. Since he had nothing to do, I invited him to fly to the United States at our expense and visit the Pennsylvania companies with us. The plan was that we all would return to Cincinnati where he would give us his thoughts about what he saw in Pennsylvania and review the information that had been obtained from the German equipment makers. He was thrilled with the invitation.

Schott/Duryea had been established as a subsidiary of Schott Glass of Germany, one of the best-known optical glass makers in the world. The U.S. government required that a major portion of its military optics be produced domestically. At the time that requirement, as well as a large nonmilitary U.S. glass optics industry, were the reasons Schott located here. With the gradual demise of the U.S. nonmilitary optics business, and cutbacks in military optics procurement, Schott had substantial excess capacity. They were eager to work with us, and it was clear their capability exceeded what would be required. We were encouraged by the visit.

Plummer Precision Optics made a wide range of lenses for military and nonmilitary applications. It was regarded by many as the best American glass lensmaker. Unfortunately, it had very little manufacturing capacity for large lenses. Our needs would have required all new equipment. Their rough cost estimates far exceeded what we required to have competitive advantage. Plummer was ruled out as a potential source.

Muhlschlag had collected a large quantity of good information. The equipment makers, DAMA and LOH, could supply us with excellent machinery for the process. Unfortunately, there were four problems; it was expensive, the process was labor intensive, lead times were long, and we had no one with the experience to operate such a factory efficiently. We concluded it was risky to proceed along this path because, even if we executed the establishment of the facility well, we still would have lens costs that were too high. In May, we still did not have a good solution for the extremely important thermal focus problem. It was a critical problem that we all felt pressured to solve soon.

# 50 A Great Answer to a Problem and a Phenomenal Deal

Within a month of the disappointing investigation into building a domestic glass lens capability, Bruce Shomler of Vivitar visited us. He was quite excited about new Opcon 35mm photographic zoom lens designs made possible by the inclusion of a few plastic aspheric elements into predominantly glass systems. We had been through this before when they tried to incorporate an asphere into a lens and remembered well the Japanese makers resistance to using elements from us. He said this time it would be different, because they planned to make these unique lenses at BASO in Taiwan where there would be no reluctance to work with us.

Of particular interest to us was a reasonably small 28mm to 200mm zoom design that would allow a photographer to use one lens for everything from extremely wide angle all the way to distant telephoto shots. What really piqued our interest was his opinion that BASO could be purchased for a reasonable price. Moreover, if USPL acquired it, Vivitar would agree to buy lenses in quantities that would utilize much of the factory's capacity. He further said that if we were interested they could help us make the deal. Although this was intriguing, he didn't know that what was even more interesting to us was using BASO as a source for TV lens glass power elements.

On June 30 I flew to Los Angeles for two days of meetings with Vivitar Chairman and CEO John Best, President Jay Katz, Chief Counsel Ken Musgrave, and Shomler. They requested the meeting to more fully explain what they had in mind and to reveal fascinating aspects of their current relationship with Robert Bosch as well as its subsidiary, Bauer & Son Optical—the company everyone referred to as BASO.

In a 1981 move to diversify its supplier base, and for the first time to free itself of total dependence on Japanese lensmakers, Vivitar had signed a five-year manufacturing agreement with BASO. The objective was for the Taiwan facility to eventually supply up to 300,000 photographic lenses per year. There were two products in production—a 28mm wide angle and a 70-200mm telephoto. The 70-200mm was a difficult design to manufacture, resulting in delivery problems as well as incoming inspection rejections at Vivitar's Santa Monica headquarters. There had been disagreements on specifications and communication interpretation that caused the relationship to be troubled. I was told that Bosch wanted out of the contract.

Releasing Bosch from the contract was a huge problem for Vivitar because BASO was where the high priority 450mm, f/4.5 MUM lens was being developed. Vivitar had already announced to the world that the lens would be available soon. Another problem was they had invested $150,000 in production tooling for the 28mm and the 70-200mm, although similar products could be supplied from Japan with performance not to the same high specifications but nevertheless acceptable. In addition to the threat of the 450mm MUM program terminating, no contract with BASO meant the end of any plans for a revolutionary 28mm to 200mm zoom and other unique designs incorporating plastic aspheres made outside of Japan.

The bottom line of the situation was the Bosch/Vivitar relationship was in irreparable shape; Bosch wanted out and Vivitar was not going to let them default without major concessions. They told me the major concessions that would be acceptable to them would be for Bosch to sell the operation to USPL at a reasonable price with the understanding we would continue their programs.

John Best and the other executives promoted this idea with me. They were correctly worried because the MUM program had dragged on for a long time through no fault of ours, and we were losing interest in it. They knew that such a deal would rekindle that interest. Given our long history with Vivitar which thus far had produced no profitable business, I probably would not have been interested in pursuing their proposal if it had not been for the PTV thermal issue. BASO, as an immediate answer to the thermal stability problems, and as a source for institutional projector lens color-correcting elements, made the consideration of such a deal extremely interesting.

Best, Katz and Shomler assured me that Vivitar would buy large quantities of the 28mm and 70-200mm. They believed the quality problems with the 70-200mm could be fixed. Musgrave said good agreements could be written that would serve both our interests. Finally they urged me to go to Taiwan, study the facility and have the BASO management completely review the details of the operation.

Upon returning to Cincinnati, our board, Dave Hinchman, Brian Welham, Jack Collins and I discussed the proposal in detail. We were all enthusiastic about what it could mean in solving the thermal problems. Also, the idea of making a major diversification into photographic optics was very attractive. We decided to aggressively investigate BASO as Vivitar had requested. Plans for a site visit were arranged immediately. Jack Collins and I departed for Taiwan July 30, with the objective of learning everything we could about the faraway company in three full days.

It had been sixteen months since Brian and I had been at BASO. It was Jack's first foreign trip and first Boeing 747 flight. We had been upgraded to first class because of an oversold business class section. He was quite pleased with the service and even more pleased when his suitcase, one of the hundreds on the plane, was the first to come out on the conveyor in Taipei. I told him this never happens, and that I could only take it as a sign of good luck as normally airport baggage waits were long with a lot of people constantly bumping into each other. A half an hour later my bag showed up and we took a taxi to the spectacular Grand Hotel.

The next morning we met BASO's Assistant General Manager in charge of operations Dr. Peter Scherb, who drove us the 100 miles south to Taichung, Taiwan's third largest city where the factory was located. The drive gave us the opportunity to get much better acquainted with Peter, learn about the Vivitar relationship and discuss factory details. BASO reported to Dr. Claus Hoffmann, the President of Bosch-Japan. Peter told us that Hoffmann had unilaterally terminated the BASO/Vivitar contract in June because Vivitar had cancelled orders for the 28mm, the one product that could easily be manufactured, leaving only the difficult, and perhaps design-flawed, 70-200mm in production. Vivitar considered Hoffmann's letter terse and not legal under the agreement. A furious John Best quickly escalated the matter to Rudolf Stahl, a top Bosch executive in charge of all foreign operations and member of the board at the Stuttgart, Germany headquarters. Stahl, we later learned, was not at all happy with the way his subordinate, Hoffman, had handled the matter and, although he agreed with the contract termination objective, wanted the matter settled in a high level manner without rancor. Bosch clearly wished to get out of the optics manufacturing business in Taiwan because they, too, had doubts about doing business with Vivitar and had been losing money. All of this was very useful information.

In addition to Austrian Peter Scherb, the remaining all-German BASO management group consisted of General Manager Wolfgang Hengst, Production Manager Peter Miosga, Engineering Manager Horst Lange, and a young accountant named Ramhorst. After a full day of history, situation overviews and discussion of the Vivitar relationship with Scherb, we were joined by Hengst for dinner where he told us the Bosch side of the story. Although a seemingly pleasant fellow, it was obvious that unlike Scherb, he was not close to the details of the operation. He reiterated that Bosch wanted out.

BASO was the first company to locate in Taichung's first trade zone in 1970. The trade zone had significant advantages that benefited the company. Zone designation made it possible to import raw materials without duty and avoid certain taxes when selling goods out of the country. Many American and Japanese companies had located in it. Canon manufactured cameras next door.

On our second day we met the management group and thoroughly toured the two plants that comprised the operation. Although I had been through them before, this time I was focused on every detail of their 35,000-foot, four-story building housed 220 people making metal mount parts, doing assembly and performing quality control functions. In addition, the offices and twenty-nine management/administrative people were located there. Across the street the lens fabrication operation was housed in a newer one-story, 43,000 square-foot building employing ninety-five people. Both operations were impressively clean and well organized. The depreciated book worth of the fixed assets was $2.5 million. We and Peter Scherb estimated the replacement cost of everything was between $5 million and $6 million. The cost to hire and train such a workforce was nearly impossible for us to estimate, but we knew it was substantial. As Jack and I moved through the facilities, we became more and more impressed with what was there and with the expertise of Peter Scherb. He had a detailed grasp of everything we viewed. Of particular significance was his conviction that the lens grinding, lapping, polishing and coating equipment could be employed to make a wide variety of large PTV elements. Nothing in the afternoon capacity, equipment or financial reviews was negative.

Jack spent our final day with the accountant going over the books, administrative and selling costs, direct labor, manufacturing overhead, raw materials, supplies, government regulations, relationship with the trade zone, banking, and other details related to our due diligence. He was shocked to learn Taiwanese law required the official accounting to be done in Chinese and that there were differences in methodology. Fortunately, BASO kept a second set of books in German that could be more easily translated into dollars and English. I had Scherb and Miosga do a few quotation exercises for me to better understand pricing. With Scherb and Hengst, I was briefed on all the key people. The entire afternoon was spent with Scherb reviewing costs and quietly discussing his views on what had to be done to make the 70-200mm zoom manufacturable. I was very comfortable with him, and had concluded he was a major talent. Jack had similar feelings. After telling him about USPL's history, current business and our aspirations for the future, we explored his interest in joining USPL as the President of BASO if we made the purchase. He was favorably disposed to consider it.

The last bit of valuable information we learned before leaving was that Robert Bosch had a tradition of never closing a factory. They would completely change product lines or sell operations, but never simply close an operation. It was part of their culture.

As we departed Taiwan for Los Angeles where we would report our findings and interest level to Vivitar, Jack and I concluded that BASO's 344 people were well trained, its 78,000 square feet of factory space was superb, the machinery and equipment were excellent, and that if we could make a deal we would at last have an answer for the PTV thermal problem. The only open questions were the manufacturability of the 70-200mm, and whether we could obtain acceptable commitments from Vivitar. BASO was working on modifications of the 70-200mm that were to be submitted three weeks later in late August. Our report to Vivitar was very positive. They continued to reassure us that they would buy both lenses at levels ranging from $3.5 million to $5.5 million annually. The 450mm MUM would be additional volume, and the 28-200mm would more than replace 70-200mm business when it was introduced to the market. John Best, Bruce Shomler and Ken Musgrave were planning to meet with Bosch's Rudolf Stahl in Stuttgart on September 22 and 23 to settle their dispute. If we were going to make an offer, all issues and agreements between USPL and Vivitar had to be resolved beforehand. With that accomplished, John Best wanted me to be with him in Stuttgart so our offer could be coupled with his negotiation.

Jack and I returned to Cincinnati to contemplate an offer, decide what would be required from Bosch, determine commitments needed from Vivitar and explore business plan scenarios with Dave, Brian and others who could provide input.

Peter Scherb and the BASO team continued to work on the 70-200mm modifications in an effort to solve the quality problem. As we built variations of a business plan, enthusiasm for the deal heightened. In anticipation, Don Keyes and I visited Schott Glass in Duryea to specifically determine blank costs and lead times for two lens designs. We were under heavy pressure from institutional projector Delta II-D customers for a thermally stable version that Brian and Opcon had developed called Delta II-DT. A new improved version of Compact Delta I, called Compact Delta 7, could be made considerably more stable with a glass element, so we also obtained quotes for it. Although the costs were reasonable, the shipping time on water to Taiwan was approximately four weeks. It was longer than we had hoped, and considering the finished elements would have to be shipped back to Cincinnati by water, we began to realize the logistical aspects of such a supply base would be very different from anything that had been experienced by our company.

Peter flew from Taiwan to Cincinnati arriving August 25 in time for dinner with Dave, Brian and me. This was the first time for Dave to meet him. With plans developing for the acquisition, it was important they get to know each other. The following day Peter met with Vivitar's Shomler and Musgrave, who came into Cincinnati to review with him and Ellis Betensky the 70-200mm modifications and progress. His presence gave Jack and me a chance to test the reasonableness of the business plan and to make a final determination of what would be required from Bosch for us to proceed. We also had him meet with our attorney, Don Lerner, who needed further information before drafting the legal conditions of an offer. After giving us their inputs, Dave and Brian departed on an earlier-scheduled trip to Japan. Peter and I flew to Los Angeles for final meetings with Vivitar product engineers and Mel Kreitzer and Jacob Moskovich of

Opcon-Cincinnati on the 70-200mm. We also finalized what would be required of Vivitar. Everything was coming together as we approached the September 22 and 23 meetings in Stuttgart.

I departed for Germany on the 19th—a day earlier than necessary for two reasons. Given the importance of the negotiation, the first was I wanted to be sure my body was well adjusted to the time change, and the second was to stop in Mainz for an afternoon meeting and dinner with Raimund Muhlschlag. He was a trusted friend who would keep in confidence the details of what was revealed to him about BASO. I was curious to know if he had any thoughts or questions that would provide additional perspective. All of his comments were favorable and supportive. From there it was on to Stuttgart the next day.

John Best, Bruce Shomler, Ken Musgrave and Peter Scherb were already there. Dr. Hoffman—the Bosch-Japan President who had a reputation for being curt and tough was also there because, as I later learned, top man Rudolf Stahl, was still irritated by the Vivitar contract termination letter and wanted him to be a constructive force in resolving what could escalate into a serious lawsuit. Hoffman had specific instructions to be pleasant and accommodating. The first evening after arriving he invited us to dinner. Best and Musgrave had previous plans, but Shomler, Scherb and I accepted.

The plan was that the next day the Vivitar people would meet with Stahl, Hoffman, Scherb and Bosch lawyers in the morning to resolve tooling and contract claims. With success in that step, Vivitar would release Bosch from the remainder of its contract if they sold BASO to USPL. I would meet with Bosch in the afternoon to make our offer.

The dinner, hosted by Dr. Hoffman, is a vivid memory. In contrast to everything I had heard, he could not have been more charming. Early into cocktails he said, "In Germany we call our good friends by their first names—would you mind if I called you Roger?

Call me Claus." I said, "Of course, Claus." Having done business in Germany for years, my experience was that even long-time good friends addressed each other formally. In any event, he was quite pleasant. Shortly after we ordered dinner the waiter brought him a huge plate of chopped raw meat. Actually it was steak tartare. The waiter then poured a liberal quantity of liquor over it, which I assume was brandy, followed by the addition of two raw eggs. Claus looked very pleased as he surveyed what was before him. Having so often heard how tough he was, I thought to myself that he had the excited look a salivating Doberman pinscher or Labrador retriever might have viewing the same scene. I couldn't resist turning to Shomler and saying, "Bruce, I certainly am glad you have to do business with him tomorrow before I do!" We had a good laugh with Claus looking slightly bewildered. On the drive back to the hotel he went over the reasons why he thought BASO should have substantial value to us.

The morning Bosch/Vivitar meeting had apparently gone well. All John Best told me was that everything was in order for me to make our offer. The afternoon meeting was attended only by Stahl, Hoffman, two lawyers and me. After the beginning niceties they asked what we were prepared to pay for BASO. I thanked them for their complete cooperation in the due diligence process and told them we did not want the offer to offend them, so I hoped they would think about it from our point of view. I said that for our small company to take over BASO represented a tremendous risk. Difficulty with the 70-200mm, new obligations to Vivitar, additional capital investment and continuing losses before a turnaround would leave us in a hole that was hard to measure. Because of this, USPL would offer one U.S. dollar with certain conditions.

The conditions were:

1. We would purchase buildings, fixed assets, inventories, supplies, tooling and miscellaneous items owned at BASO. Bosch would retain all liabilities.

2.  Bosch would keep the inventory level of raw material, work in process and finished goods no less than the book value as of June 30, 1983 until closing.

3.  Bosch would furnish the services of Dr. Scherb, Mr. Miosga and Mr. Lange at its expense for eighteen months from a December 31, 1983 closing date, and the services of Mr. Weisser and Ramhorst for six months. This included both the German and Taiwanese portion of their compensation.

4.  Accrued vacations and holiday pay as well as Chinese New Year bonuses would be paid to December 31.

5.  USPL would form a new company called BASO Precision Optics Limited that would assume all Taiwan employment liability after the closing date.

There was a long pause before Stahl spoke. He thanked me for doing our investigation in Taiwan and coming to Stuttgart. He said they would like to talk about it for awhile. Lastly he invited me to dinner with them and the Vivitar people. I accepted and departed the meeting, very uncertain as to what to expect.

That evening, everyone was very gracious during cocktails and an elegant dinner hosted by Stahl. There was no business conversation. They did ask to meet with me the next day at mid-morning. Hoffman was given the job of driving me back to the hotel. On the way he said, "Roger, can't you really pay more than $1.00 for such a valuable property?" I had the perfect answer. I said, "Claus, Robert Bosch is one of the great companies of the world. If it, with all your expertise and resources, couldn't be successful with BASO, it was very presumptuous and risky for us to even be offering $1.00." He had tried for something better but didn't press the point. I thought my answer made sense to him.

The morning meeting with Stahl, Hoffman and the lawyers took place after they met to decide what they wanted to do. Stahl said they would accept all of our terms but one. For legal reasons relating to the complex way the ownership of BASO was held by Bosch and various subsidiary companies, they needed us to buy the corporation instead of just the assets. They said an asset deal was extremely cumbersome for them—maybe not even doable. This meant we would have to assume all the stated liabilities and, perhaps just as important, all unknown liabilities that may have existed. I said that if, in fact, it was just a legal problem, we would buy the corporation with a purchase agreement that called for the practical effect of the deal on us to be the same as an asset-only purchase, and we wanted it guaranteed by the parent Bosch Corporation. I did this without advice of counsel but with the confidence we were dealing with highly reputable people who were part of a superb company. They agreed that, with the full backing of Bosch, for us it would be like an asset-only purchase for $1.00. We had a deal!

I was ecstatic. We had just bought assets worth $5 million to $6 million for $1.00 and acquired an excellent low-wage workforce that would have cost millions more to assemble and train. I called Cincinnati with the news. Everyone thought it was an incredible deal. Someone said, "We will send you another $5.00—go buy some more companies while you are there!"

It would have never happened without the thoughtful opinions as to what might be possible that had been solicited from Peter Scherb throughout the process. Although he was always loyal to Bosch, his openness and encouragement gave me the confidence to press forward.

Vivitar agreed to buy 6,300 of the 70-200mm at $70 each, and 3,600 of the 28mm at $30 each in 1984, for a total sales value of $5,490,000. They also agreed to at least the same level of purchases for 1985. Much of the finished parts inventory for the first year's production had already been produced and was part of our $1.00 purchase.

The Stuttgart agreements gave everyone what they wanted. Bosch was released from a contract and was able to exit a business they very much wanted out of in a way that, while costly, was not too costly for a multi-billion dollar company. And, they did it in a manner that preserved their important cultural value of never closing a factory. Vivitar's top priority for BASO was the 450mm MUM lens followed by the possibility of a breakthrough 28-200mm zoom development. Both depended on USPL's continued interest in working with them. The incentive was to help us buy BASO at an unbelievable price and the promise of substantial purchases of the two existing BASO lenses. Neither Vivitar or Bosch knew our primary interest was the acquisition of a cost-efficient factory capable of making power elements for PTV lenses that would solve the thermal stability problem threatening the majority of our sales volume. The prospect of substantial Vivitar business was attractive to us, but if BASO had not had the capability to make the TV lenses we would not have pursued the deal.

In the three months to the December 31 closing, a very complicated sale agreement had to be written because of the guarantees Bosch made as a result of our accord. Since we were buying the corporation—not simply starting a new one—we had to understand all details of the books, contracts and previous obligations. Jack Collins, who had been immersed in the deal since its inception, and our lawyer, Don Lerner, went to Taiwan to review everything in detail. Don found a Chinese law firm with excellent credentials for the legal review. Jack engaged the Taipei office of Arthur Andersen to verify that the books, kept in Chinese, were in proper order, the tax returns had been accurately filed, and to perform an audit.

For Bosch, the deal had become complex because they were selling a corporation with ownership residing in the German parent and subsidiaries in Switzerland, Luxembourg and Malaysia. We had no idea why it was held that way. We did learn that each of these countries had its own peculiar laws that required legal effort to make the sale. It was important Don Lerner understand those issues. We had our own problems. Because USPL was a Sub Chapter S corporation, it could not own more than 80% of a foreign corporation. To comply with that rule USPL would own slightly less than 80%, with the balance being held personally by its shareholders on a pro rata basis.

Given the tangled maze of different languages, laws and accounting methodology, it was remarkable that everything that had to be accomplished took just two months. Jack and Don, along with the Bosch lawyers, had done a spectacular job of making it all happen. On November 12 Dave, Jack and I flew to Taiwan for the signing of the sale agreement. It was the first time Dave had seen BASO and its people other than Peter Scherb. He was impressed.

The *pro forma* guaranteed balance sheet that would be adjusted December 31 contained $2.5 million of fixed assets, $1.1 million of inventory and $245,000 of cash. Bosch had to make the hard assets equal the liabilities and shareholder equity because they had insisted on the sale of the corporation rather than just the fixed assets and inventory. Considerable receivables which they believed were collectable were not shown, but the statement footnotes and agreements called for them to be guaranteed. The bottom line was that for Bosch to make it a corporate sale, they had to inject over $1 million in cash. Months later, for reasons involving the nature of their ownership that I never understood, they had to put in somewhere between $500,000 and $750,000 more cash. The cost of not doing an asset sale had become substantial for them.

After the signing, we celebrated with the key Chinese managers and the Germans that were staying on for at least eighteen months. It brought renewed hope for the future of BASO to the Chinese employees and potential career options for the Germans as well as their Austrian leader, Scherb, if BASO could be made into a success. Everyone had a feeling of optimism.

Dave and I stopped for three days in Japan on the way home to tell the Japanese customers about our acquisition and what it meant to solving the thermal problem as well as improving chromatic aberrations through color-corrected designs. Meetings were held with Sony, Hitachi and NEC. These were the companies that had been exhibiting the greatest concern. They were delighted. As soon as we arrived home RCA, Magnavox, Matsushita, Toshiba, G.E., Zenith, Barco, Philips and Thomson were given the news. Everyone thought it was an extremely important development for the future of USPL lens technology.

# 51 1983

As expected, fiscal year 1983, ending June 30, was a disappointing step backwards. Given the severe adjustments in operations that had to be made, I thought our management team performed magnificently as was shown in the financial numbers. The first half of the year (July 1–December 31, 1982) resulted in a slow fall season and poor Christmas sales, causing manufacturers to enter January with huge inventories. This in turn caused further reductions in orders and, in some cases, delay of new products utilizing Compact Delta I. The poor market conditions required that expenditures be cut throughout the company. The biggest savings came from cutbacks of people, many of whom were temporary employees who had been hired during the last big manufacturing buildup. The results of the scale-back were impressive.

At $14,600,000 sales were down 33% from fiscal 1982s $21,850,000. It was the first time we had experienced a sales drop. Net profit, after a full corporate tax provision, fell to $1,743,000—down 20%. To absorb a 33% fall in revenue, and have after-tax profits actually increase from 9.9% to 11.9% of sales, was a result for which we all could be very proud. Most businesses that experience sudden severe reductions in sales have their profits obliterated.

Standard optics division external business grew 16% to nearly $3,000,000, while its internal lens housing volume for the massive-optics division fell by 45%. This meant the associates in that part of the company also were affected by the TV turmoil.

The second half of calendar 1983 (first half of fiscal 1984) brought a sales rebound as TV makers cleared the excess inventories of old models and introduced sets with Compact Delta I. This had been expected, so the company was well prepared for a production ramp.

The economic policies of the Reagan administration and the Federal Reserve Board led by Chairman Paul Volcker were continuing to have a positive effect. Volcker's number one priority was to curb inflation which had reached a devastating high of 13.5% in 1980. In 1983, it was down to 3.2%. The prime bank lending rate was more stable than it had been in recent years; although down over 5% from the previous year's peak, it still was high ranging from 10.5% to 11.2%. Unemployment remained at a high 9.6%. People were beginning to spend more freely as reflected in a drop in the personal savings rate from 9.0% to 6.7%. Of continuing concern to us was the undervalued Japanese yen which fluctuated between 227 and 247 to the dollar. Surprisingly naive comments by Secretary of the Treasury Donald Regan indicated he had little understanding of how difficult the yen-dollar currency relationship made it for Americans to export and how easy it made it for the Japanese. Currency fluctuations in effect were pricing fluctuations on foreign sales, which were a large part of USPL's volume. It was a subject we worried about very much.

The Compact Delta I allowed TV makers to enhance significantly the form factor of their PTV sets. In spite of this advance, cabinets continued to have a front-to-back dimension that we thought was excessive. Also, the lens was expensive. In January Brian Welham, Ellis Betensky and I went to my condominium in Boca Grande, Florida for a two-day session on what we might do that would permit sets to be made smaller, better and less expensively. It was a highly productive meeting, during which the concept was developed for a new lens that would reduce both cabinet size and price. We called it Compact Delta 7. The lens would have a shorter focal length, smaller elements and a better light bending angle. A key requirement in the design was for us to be able to manufacture rectangular elements which would minimize the separation of the three CRTs. The lenses would be offered in air-coupled and liquid or optically-coupled forms that improved contrast. Knowing that we eventually would need to use a glass element for thermal stability, such a provision was included in the design requirements.

Opcon went to work on the design and we initiated a crash development effort to make rectangular elements. Both were accomplished successfully and quickly. With the design in hand, cabinet configurations could be mocked up that demonstrated the advancement. It was dramatic. Our customers, many of which were just introducing Compact Delta I models, were excited about the improvement. The only unfortunate aspect of the program was that a design change after only one year required that both USPL and the TV makers amortize existing expensive tooling over a shorter life span than was optimal. The compelling advantages caused us to move forward immediately. Allowing time for prototyping and pilot runs on the part of the manufacturers, the lens was scheduled to go into full production nearly a year later in April 1984. It would again ratchet the technology upward.

Delta II-D was utilized by Sony and others for institutional two-piece projectors. Sony, in particular, was unhappy with the lens' thermal stability. A new design, called Delta II-DT, was hurried into production late in the year. The T stood for thermally-corrected. It contained the first glass power element obtained from BASO. The significant improvement in focus stability that was achieved with the modification delighted Sony. BASO was benefiting us even before the deal was closed with Bosch at year-end.

The outlook for 1984 was excellent.

# 52 BASO: The First Year

Soon after the Taiwan acquisition, we noticed a marked change in the attitude of PTV lens customers towards USPL. Our new ability to supply glass-plastic hybrids eliminated everyone's concern about the thermal flaw. Some believed the lens color-correcting capability BASO brought to us was equally important. It was as if a huge load had been lifted from our shoulders.

Getting off to a good start with Vivitar was a greater challenge than we had anticipated. The 70mm-200mm zoom had a high reject and rework rate because it failed a drop test during their incoming inspection. It took a little over two months to correct the problem. The 28mm was no problem, but Vivitar was slow to release orders for both items. Our agreement as to what they would buy was never put in contract form and signed. We thought their word was good and the 450mm MUM development gave us all the leverage that was needed.

In March of 1984, John Best complained at length to me about pricing pressures in the photographic lens market. He said they were buying an acceptable 28mm from a Japanese supplier for $26, and one of his customers was paying $25 for a competitive lens in lots of 10,000 delivered to Dallas and Denver. Our price was $31.50. Although our 28mm was acknowledged to be better, quality was not given much weight in John's mind. Also, he said they were selling a Japanese supplied 70-210mm to their customers for $79 while paying us $74 for the 70-200mm. Again, there was no disagreement on the fact our quality was superior.

It sounded like the beginning of a price negotiation to me. We wondered if they would have been getting those prices from the Japanese lensmakers if USPL had not bought BASO. There was no doubt in any of our minds that Vivitar's Tokyo-based executive was using the event to push for more favorable deals. In April I went to the Photographic Manufacturers Association Show in Las Vegas primarily to meet with Best and Shomler about the lack of shipments they were calling for from Taiwan. I also told them we lost over $250,000 at BASO in the first quarter. This was a correct number based on a full product costing. In reality, the situation for us was not so bleak because of the large inventory and cash account that were purchased for $1.00. Furthermore, glass PTV lenses were being sold to USPL at very favorable transfer pricing. Nevertheless, with some concern, we looked beyond the fortuitous benefit of having a $1.00 inventory because it was not going to last forever.

Best and Shomler were sympathetic to our problem, but not supportive or encouraging in a tangible way. The Photographic Manufacturers Association Show was not going well for anyone, and there seemed much uncertainty about the state of the photographic equipment business. I pressed them to honor the agreement and, while sympathetic to their situation, considered it a problem they had to solve. As frequently was the case with Vivitar, nothing was firmly resolved. Again, I wondered if it was worthwhile trying to do business with them.

BASO's supply of glass elements to USPL for thermal versions of Delta lenses went smoothly and more than met our expectations. We continued to muddle along at Vivitar with shipments far lower than had been anticipated. Dave Hinchman became very involved in the administration of the relationship—particularly the development of the 450mm MUM lens. Brian Welham helped on the technical matters. In May, Dave told their product development manager, after much consultation with Peter Scherb, that the price from our Taiwan factory would be $274 for the first 2,500 units and then drop to $247. The manager was in shock. They had floated retail numbers to the trade in the area of $350; $274, with two distribution channel markups to be added, required a major change in thinking. We told them they had designed a Cadillac or Daimler—not the value lens that had been touted to the world. Redesign attempts to cut costs would further delay the product and the revenue stream on which we had planned.

Peter Scherb and his German management team were very engaged on our behalf even though they were to remain on the Bosch payroll to mid-1985. Communications were not a problem. Jack Collins was responsible for administrative and financial details in Taiwan and quickly developed a good working relationship with the Chinese accounting manager. Our only real problem was the order levels from Vivitar and redesign delays with the 450mm MUM. The MUM project took an inordinate amount of Peter's time.

In Taiwan and China, legal signatures are performed with a chop. It is a small piece of wood that has carved into it personalized Chinese characters and design on one end that is used much like a rubber stamp with which Americans are familiar. The chop is first tapped on an ink pad and then pressed against documents requiring signatures. Much to my surprise, we learned that as Chairman and CEO of BASO, I had a chop that was used by the Chinese managers whenever necessary. Peter Scherb had one, too. As a person from a country that places considerable emphasis on careful management controls, it was a very disturbing thought that people I barely knew could legally sign my name to whatever they chose and there would be a binding agreement. After much discussion about it, we asked that I be notified of all documents on which it was used. The USPL Board indemnified Peter and me against unauthorized use. There was never a problem with the chops, and I quickly became comfortable with the trustworthiness of our Chinese employees.

Peter informed me that the Chinese employees had given me, as was their custom, a Chinese name. It was Lo Chi Hao. The translation was:

Lo—to arrange over a wide space

Chi—lucky

Hao—outstanding in intelligence and talent.

Not wishing to offend anyone, I did not inform them they had been overly generous with their creation. Besides, I fantasized that they may have known something I did not. One of the employees, Mrs. Ching Tse Cheng, painted an oriental scene and, in Chinese calligraphy, wrote my new name. It was presented to me in an attractive frame. Having, on a few occasions, been referred to less charitably, it was a gift I very much appreciated receiving.

The BASO workforce was impressive and, for the most part, experienced. Because the company paid higher wages than the average and had been the first to build in the trade zone, there were many long-tenured employees. The wages, on a monthly basis, ranged from $130 to $165.

By November, our business relationship with Vivitar had seriously deteriorated. Although regular shipments from BASO had continued, the quantities were less than had been agreed upon. Dave and I visited John Best and Bruce Shomler because we couldn't continue along the path we were on. The situation we found in Los Angeles was discouraging. The day we were there they reduced the organization by forty people—many at high levels. A public offering of their stock had to be cancelled because of the state of the photographic lens business. In his trip report, Dave described our options as "bad, worse and worse yet."

They claimed to have inventory that would last at least six months. There was no interest in talking about anything beyond that except the long-suffering 450mm MUM project. In response to discussion about their commitments, Best said he felt some obligation but offered no relief. Dave felt Shomler did not even feel the obligation. We told them we had lost $1 million at BASO to date, and something had to change. We, in fact, did have a $1 million book loss, but since most of the parts in the products which had been shipped were from the $1.00 inventory, we actually had a positive cash flow. And USPL was buying large quantities of PTV glass elements at extremely favorable prices.

It was clear our hope for Vivitar becoming a large profitable customer had come to an end. The two lenses we were making for them had no future. The revolutionary 28mm-200mm had fallen as a priority. We reluctantly agreed to support the 450mm MUM because they reluctantly agreed to buy enough of the two existing products to consume the work-in-process and finished goods inventory.

In less than a year from the acquisition, we were in the earliest stages of disengaging from Vivitar. In spite of that, the primary reason we bought BASO was as valid as ever. It gave us instant credibility with the PTV makers and an immediate source of glass lens supply.

A new strategy and business plan had to be created. By January 1985, Peter, Dave, Jack and I were exploring the options. One was to try to sell the BASO capacity to other lens buyers. This was rejected because there were many companies in Japan and other countries that could make similar optics. This meant highly competitive pricing with nothing special being provided by us. Another was to assemble two of the volume PTV lenses that were being shipped to Japan. It would have required shipping plastic elements and housings to Taiwan to be combined with the glass elements. That proposal did not compute well either.

Eventually, we came up with a third idea that did make sense. Because the operation was neatly divided between two buildings, we could attempt to find a buyer for the older four-story, 35,000 square-foot metal mount plant that had 220 employees. Roughly twenty of them, involved in accounting and administrative functions, would be transferred across the street to the newer 43,000 square foot glass lens processing factory. The goal would be to find a buyer that could utilize the metal plant and its workforce.

We became more and more comfortable with Peter Scherb in the year-and-a-half of our association. His technical background and work experiences were impressive. In 1974, he received a Doctorate in Physics with honors from the Technical University of Vienna. After a year as a field engineer with Schlumberger in Australia and Indonesia, he moved to German camera maker Rollei in Singapore working as a manufacturing manager, and then as head of quality control. In 1977, he was hired by Bosch to be the technical manager of BASO. When we met him, he was in charge of all operations. Peter's high standards for the organizations and his own performance were obvious. The clean plants were well laid out for optimal work flow. My sense was that he was the primary driver behind what had been acquired.

If the metal mount plant was sold and we exited the photographic lens business, there would not be enough remaining activity for the top job to be attractive to Peter. I started discussions with him in January about the possibility of his coming to Cincinnati as manager of production technology. It was an area of USPL that needed continuous improvement, and we were convinced that he was qualified to be a major contributor. There was another reason why I thought it would be good to have him at USPL. Because of the long shipping time to and from Taiwan, I was beginning to ask myself whether we also needed to have a glass lensmaking capability in Cincinnati. He had the experience for leading an effort to create a low manufacturing cost domestic factory if it could be done.

Assuming the metal mount plant was sold, and Peter moved to the U.S., we would then be left with the need for someone to run the remaining Taiwan operation. Peter Miosga, the German manager of the glass lens plant, was the logical candidate. We thought that with Miosga reporting to Scherb and the Chinese administrative/accounting manger reporting to Collins the structure would be workable, since without the metal mount plant it would be far less complex to manage. Over the next couple of months, we continued to discuss the idea.

*August 2, 1978 Norio Itoh,
of Sony, visited USPL, marking
the beginning of our relationships
with the giant Japanese electronic
companies.*

*Phil Jones, Manager
of the Ohio Trade
Office in Tokyo,
with Roger Howe,
a Japanese customer
and Brian Welham
during one of our first
trips to the Orient.*

*Sony coffee table
projection TV intro-
duced in the summer
of 1980, utilizing
Delta II lenses.*

— 277 —

*Jerry Behne joined USPL as head of Human Resources in 1970.*

*Zenith "pop-up" projector introduced in 1981 between conventional rear screen PTV's.*

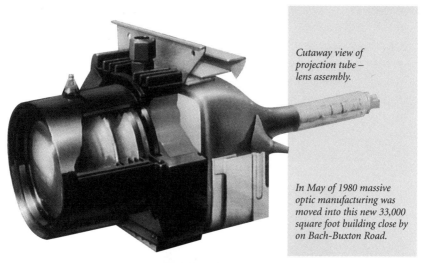

*Cutaway view of projection tube – lens assembly.*

*In May of 1980 massive optic manufacturing was moved into this new 33,000 square foot building close by on Bach-Buxton Road.*

u.s precision lens

*Jack Collins joined USPL in 1981 as Chief Financial Officer.*

*Dr. Peter Scherb joined USPL January 1, 1984 as President of its Taiwan BASO subsidiary company.*

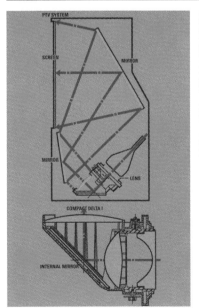

*USPL Projection Television demonstration room.*

*Compact Delta I schematic showing how image path is turned 90 degrees by an internal mirror in the lens.*

*Compact Delta 1.*

— 279 —

*Brian Welham giving the "no advance notice" seminar to 15 Chinese engineers on making PTV at Radio Factory #32.*

*Madam Tu, the 34 year old plant manager of 1,500 employee Radio Factory #32 in Shanghai, China with her staff and R. Howe.*

*January 1, 1984 USPL acquired BASO Precision Optics located in Taichung, Taiwan from Robert Bosch of Stuttgart, Germany for $1.00.*

# Part Four

*Transitions*

# 53 1984

July 1983 marked the beginning of fiscal 1984. TV manufacturers had disposed of virtually all old model PTV's. With inventories low as most makers introduced models with Compact Delta I, our business was strong throughout the year. Sales at the June 30, 1984 close were $20,688,000—up 42%, and profit, after a full federal and state tax provision, was $2,857,000—up 60%. It was both a record in total dollars and, at 13.8, as a percent of sales. Capital expenditures were $2,100,000 and after provisions for additional working capital, cash, net of personal tax distributions, rose to nearly $8,000,000. The good news was we were seeing the productivity benefits that come from continual running of a standard high volume product. The bad news was that 53% of our sales were concentrated with two customers —Hitachi and Sony. Hitachi sold its own brand and also provided RCA with its product.

The standard optics Honeywell auto-focus business took a nice jump to $1.2 million with the introduction of new products. Stretch Hoff, with the help of Don Keyes and others, was developing a fiber optic connector business. Hewlett-Packard was the biggest customer, and other prospects were having frequent discussions with him. It was an area we thought had very high potential.

The unlevel playing field caused by an undervalued yen, was a never ending concern. It varied from 223 to 252 against the dollar. The prime bank lending rate went up slightly, ranging from 11% to 13%. Inflation increased 1.1% to 4.3%. Unemployment dropped from 9.6% to 7.5%, and Americans increased their savings rate from 6.7% to 8.6%. None of these statistics concerned us very much, except for the dollar/yen relationship. We were constantly told about potential competitive all-glass PTV lenses that, except for the currency disparity, would have sold for much higher prices.

In April we started production of Compact Delta 7. It was a superb design, and was offered in both all-plastic air-coupled and liquid-coupled variations. There were also thermally stabilized versions with glass power elements. Comparing costs to Compact Delta I, for a customer buying 2,000 air-coupled kits of three lens assemblies per month, the price of each kit dropped $45, from $194 to $149. Thermal versions were $18 more. As soon as possible TV manufacturers converted, and Compact Delta 7 became the high production lens. In addition to the cost advantage, cabinets could be again greatly reduced in size, thereby providing consumers a less intrusive product for their homes.

In May, Jerry Behne was running advertisements in the local papers for a production supervisor. Specifically, the job he was attempting to fill was consumer PTV manager of assembly operations. The ad was answered by David E. Szkutak, a Massachusetts native who was employed at the time by Procter & Gamble as a manufacturing manager. He was a chemical engineering graduate of Worcester Polytechnic Institute and had been with P&G for five years. Jerry recognized Dave as a person with high potential for advancement and recruited him aggressively. After a few meetings he agreed to join us as the assembly leader in the Bach-Buxton Road factory.

He excelled in that position, and soon was given a production management job in the video optics special products department. This group made high tolerance institutional PTV lenses. Again, he excelled. At night and on weekends, he studied for and eventually received an M.B.A. from Xavier University. In 1986 he moved into sales engineering and was assigned to the video optics sales department. He was increasingly recognized by Dave Hinchman, Jerry Behne, me and others as a talent. In the years to come he would become vice-president of sales and marketing, video optics in 1990, executive vice-president in 1997, and president, chief operating officer in 1998. In July of 1999 he became chief executive officer. Dave Szkutak, over the years, has been an outstanding contributor to the success of USPL.

From the mid-1970s we believed it was an obligation of USPL to be a good corporate citizen. The company participated in United Appeal campaigns and later those conducted by the Cincinnati Fine Arts Fund. Because each brought benefits to wide areas of the community, all other solicitations were excluded. With particular emphasis on the United Appeal, both were heavily promoted to all associates for their participation. In Clermont County, the company and its employees have long been recognized as generous civic contributors.

In 1984, Dave Hinchman was the chairman for the Clermont and Brown Counties United Appeal campaign. With Stretch Hoff co-chairing the public employees division, Jerry Behne leading the campaign training, and accountant Pat Cowherd tracking the progress, USPL was deeply involved in doing its part. Dave and his team did a superb job breaking all records for the two counties. Internally Pete Cutler, led the USPL effort with a committee from various areas of the business. They exceeded the goal by 50%, with 85% of the associates participating. 77% were at what United Appeal called fair share levels. USPL's generosity was a significant factor in Dave's overall campaign results.

The explosive growth potential of PTV required frequent capacity reviews and plan revisions. In looking at expansion options for the year ahead, a goal was set to be able to accommodate 40% growth a year. It was assumed the actual number would vary between zero and 30%. It was a great business but highly volatile, as had been seen. With a twelve-month horizon, we wanted capacity in excess of 20% of the average month and 5% of the peak month. We were able to make adjustments fairly rapidly in all areas except factory space; that had a much longer lead time. To provide a space cushion, in November construction was started on a 72,000 square-foot building directly across McMann Road from the headquarters. Miller-Valentine, the contractor for the Bach-Buxton Road plant, was selected to do the job. The occupancy goal was early fall of 1985.

At the end of calendar year 1984, PTV sales to dealers were 195,000—up 35%. There had been no high volume liquidations and prices remained relatively stable. Since 1980, 600,000 reasonably good projectors had been sold. When 1,000,000 VCRs had finally been sold, the growth rate jumped significantly. We hoped the same thing would happen with PTV.

# 54 Intellectual Property and Patent Challenges

As mentioned in Chapter 7, in 1972, when the company was becoming better known for plastic optic capability, I was concerned about competitive recruiting efforts. If another company wanted to enter the business, it would have been logical to target USPL employees for hire. Our small size meant that key people had a good understanding of all areas of the technology. Optical moldmaking techniques, the finishing of lens mold inserts, molding machine specifications and molding cycle programming added to information that would be extremely valuable to a competitor. To protect the value of this knowledge, secrecy and noncompete agreements were signed by all associates. In addition to holding USPL proprietary information in confidence, the agreements called for a three-year noncompete period after an associate's departure.

The agreement was drafted by Cincinnati patent law firm Wood, Herron & Evans. A few years later when the massive-optics process was developed, the agreements were strengthened with specific language regarding trade secrets. While the noncompete portion was for three years, there was no time limit on keeping trade secrets in confidence. Had USPL been located in Texas or California these agreements might easily have been challenged, but Ohio had excellent laws protecting such property so the documents were meaningful.

Our policy of not showing customers or other visitors areas of the factories that we deemed to have proprietary importance served us well. I believed that even if a company could, for the most part, figure out what we were doing, in their eyes it would still be a big risk to compete unless they knew for certain the specifics. The combination of the associate noncompete agreements and barring visitors from critical areas protected us from ever having a serious loss of technology know-how.

When Ellis Betensky designed the uniquely advantageous three-element Delta II PTV lens, a patent was immediately applied for because competitors would be able to reverse engineer the configuration to understand what had been done. The fact that making it required the use of the proprietary massive-optics manufacturing process gave us an additional large measure of protection. By 1985 we had been issued four lens design patents. Over the years, these would be added to and improved upon. Eventually, by 1998, the company had a total of thirty-two patents. It would be a considerable amount of intellectual property.

In 1982 we learned that the Japanese company, Fujinon, the optics arm of Fuji Photo Optical, was about to sell an all-plastic aspheric PTV lens to Sanyo, who in turn would use it in projectors sold to Sears for resale in the United States. A customer obtained a sample for us that we analyzed to determine if it encroached on our patents. The lens had a higher f number, causing it to transmit less light than a Delta II, and our experts did not think the picture quality was exceptional. We did believe it violated our patents, causing us to send letters to Sanyo and Sears putting them on notice that we would aggressively defend our claims. The Sears product was not introduced. We never learned if it was our letter or another reason that caused the program not to go forward.

In 1984, we were about to do substantial business with the Japanese giant Matsushita Electric's U.S. subsidiary in Chicago that made Panasonic and Quasar PTVs. Dick Kraft, the subsidiary's president, was a friend who had pushed hard to get USPL optics designed into their products. In Japan, Matsushita announced the introduction of an all-plastic aspheric PTV lens and put considerable pressure on the U.S. subsidiary to design it into their products. The lens announcement included significant design detail. We were confident that it violated our patents.

Ellis had worked for a number of years on photographic lens patents with Robert Montgomery, a patent attorney in New Haven, Connecticut. Ellis taught Bob about lens design and Bob taught Ellis about the intricacies of filing for patents. Because of his expertise, we engaged Bob to represent our interests. There was no doubt in his mind that Matsushita was in violation.

I went to see Dick Kraft and told him this was a real dilemma for us. His company was about to become a large customer of USPL and we were confident it was encroaching on our patents. The problem was that if we didn't defend our position others might also encroach, yet we didn't want to cause bad feelings with him and his associates after we had worked so hard to win their business. Kraft's handling of the issue couldn't have pleased me more. He said he realized it wasn't personal and that USPL had to do what was necessary to protect its interests. So that he could better understand the issue, he asked to have Bob Montgomery meet with his patent attorney to go over the potential claim violations. We agreed that neither of us would surprise the other with rash actions.

Kraft gave us one of the Matsushita lenses to evaluate. We were pleasantly surprised because it lacked performance and was awkwardly large. As agreed, the lawyers met. Nothing more was heard about their using the lens. Again, we did not learn whether they thought we had a good case or simply concluded the lens was not adequate for their use. In any event, our careful handling of the matter and Kraft's positive reaction prevented any wounds from being inflicted to the relationship. However, we now had experienced threats from two extremely able companies that wanted to get into the PTV optics business with aspheric plastic lenses, and we expected others would attempt to do so too.

In late 1984, Fujinon announced the introduction of a lens that was nearly an exact copy of Compact Delta 7. This caused us great concern. I met with two patent attorneys at Wood, Herron & Evans to determine options for defending our position and to learn what it might cost. There were two approaches. The first was to press our case against violators in federal courts. The disadvantage was it could take years and cost millions of dollars. The advantage was that if we won, significant damages could be collected. The second approach was to take our case to a U.S. agency called the International Trade Commission (ITC). This would be possible because all the threats were from foreign companies. This route had the advantage of speed because once a proceeding began, it had to be concluded in one year. It was extremely onerous for a violator because if we prevailed, all the offenders or offenders customers' products could be excluded from the U.S. market. In one of these cases, a small U.S. auto parts maker had actually stopped General Motors from bringing in cars from one of its foreign subsidiaries. It also was far more cost effective. The estimates were that a proceeding would cost between $200,000 and $500,000. The disadvantage of an ITC action was that if we prevailed no monetary damages would be awarded. Its teeth came exclusively in the denial of access to the U.S. market.

It was an easy decision for us. If we had to initiate an infringement case it would be through the International Trade Commission. The ITC tended to be favorable to U.S. companies and was a strong defender of patents. Just the threat of such an action posed a huge risk to a violator who depended on the U.S. market. It was a mechanism that allowed a small company like USPL to defend its interests effectively against a mammoth company. We made it known to all customers that our patents were being challenged, and that we would aggressively use the ITC to resolve issues. We explained nicely why we would have to do it and what it would mean to them if they were using competitive optics that were in violation. Everyone understood the severity of the penalty and why we would have to pursue such a resolution.

Dave Hinchman took the lead in coordinating the administration of USPL patent filings. Ellis Betensky reviewed all new Japanese patent applications involving optics to make sure we were not violating other people's intellectual property and to monitor their development activity. The Opcon group and Brian Welham were very involved in the process. Legal work for all our filings was done by Bob Montgomery. This activity had to be conducted carefully and consumed quite a bit of time because a mistake or omission could have been very costly. Their excellent work resulted in USPL having a web of PTV lens patent claims that would be extremely difficult for a competitor to circumvent.

The Fujinon copy of Compact Delta 7 convinced us this was a competitor that was going to challenge aggressively USPL market dominance. February 6, 1985, Dave and I visited their manager of optical design, manager of the international department and the engineering manager in charge of patents in Omiya, Japan. After a cordial reception, we explained that we had tested some of their lenses and seen their brochure for the Compact lens. In our view, they were in violation of two of our patents and numerous claims in patents that were pending. They were told also other Japanese lensmakers were violating our patents, and that USPL's legal counsel had advised us we must take action to protect our interests. Mr. Honda, the manager of optical design, conversed in excellent English and was obviously very bright. He explained why he believed they had designed around our claims and why he thought some of them were invalid because of prior art. In as nonoffensive way as possible, we disagreed with his reasoning.

It was clear they wished to resolve the issue without conflict. We agreed to study each other's positions and meet again for further discussions. Honda said they would consider sending us a sample of their Compact lens. We agreed to send them copies of our new patents as they issued. There was no mention of licensing. Another meeting was held during the year in which we expanded on why we thought they were in violation and more information was exchanged.

Dave and Brian had a final meeting with them in January of 1986. Sanyo, Fujinon's customer, discontinued its PTV program soon thereafter. They had come to realize that we had a strong position and, more importantly, would resort to drastic action to enforce patent rights. To our surprise, Fujinon was never a serious supplier to PTV makers and eventually ceased its projection lens development efforts.

Hitachi had been our largest customer in 1984 and 1985 because it made the RCA PTVs. Rather than utilizing Compact Delta 7 in 1984, their engineers decided to pursue a short focal-length straight lens for a cabinet configuration that would have just one image-folding internal mirror. At their request, we designed and prototyped Delta 12. It was considerably less expensive and performed as they had hoped. It was used in the 1985 models.

In the spring of 1986 Mitsuharu Akatsu, who headed the Hitachi television business, informed us they had designed a lens similar to Delta 12 and were in the process of creating a manufacturing facility capable of limited volume. We were shocked. Our largest customer was about to compete with us. A few months later Mr. Akatsu proposed that we cooperate since Hitachi wished to have an important continuing relationship with USPL. They had also obtained a design patent, and although we thought they were violating ours we were not totally confident. These very able engineers would not have done what they did if they thought there was going to be a problem. We exchanged design information and studied each other's positions further. In February of 1987, I informed Mr. Akatsu that, although we believed our patents prevailed, we were willing to consider an agreement that assured us of at least half of Hitachi's volume. It would have to contain a provision in which they agreed not to sell their lenses to any third parties. It was possible they had claims in their patents that could be reasonably argued. Rather than getting into a legal debate with our biggest customer, we agreed to an arrangement with terms in accordance with what I outlined to them.

USPL and Hitachi's relationship has been excellent in all the years since then, and they have remained a large customer.

There have continued to be patent challenges, but we have thus far always prevailed. If we had not been diligent in protecting our intellectual property, USPL would have experienced a very different competitive landscape with the Japanese lensmakers.

# 55 A Shift in Manufacturing Strategy

To serve Japanese customers well, the logistical challenges of supplying glass lens elements from BASO were formidable. For instance, institutional projectors delivered to the United States made with glass lenses from Taiwan required the elements to be shipped over water four times. Each step of the manufacturing process demanded lead times that were difficult for both USPL and the manufacturers in Japan to work within.

An order of glass lens blanks requested by BASO from Schott Duryea in Pennsylvania, or from Hoya in Japan, required a manufacturing time of about four weeks and a shipping time of about four weeks by a sea carrier. Processing time in Taiwan was about four weeks. Finished lens shipping time to Cincinnati was another four weeks. Assembly took an average of four weeks, and the finished multi-element lenses required four more weeks for delivery to Japan. Assuming it all took four weeks for the manufacturer there to assemble the lenses in projectors, and four weeks to ship the final products to America for sale, the entire process took thirty-two weeks. In reality, of course, the time required for each step varied a little more or less than four weeks. However, when the work in-process and finished goods inventory cushions were added in, the lead time would well exceed thirty-two weeks.

We typically operated with far shorter lead times for all-plastic lenses. It was nearly impossible for TV makers to project exact order levels nine months to a year in advance, yet the logistics of Taiwan glass element supply required that USPL press for such lead times. We understood their problem and they tried to understand ours.

In the spring of 1985, Peter Scherb and I began to talk about the issue. Glass lensmaking had not changed technologically in significant ways for decades. It was extremely labor intensive. For that reason Japan was able to develop a highly competitive industry when it was a low labor cost country in the 1950s and 1960s. Later, other countries, such as Korea, Taiwan, Thailand and Singapore became important production sites as Japan's labor costs rose. The Europeans and Americans lost most of their once-large lensmaking industries because low cost labor was the key to being competitive. What I pressed Peter to think about was how could we reduce the labor content through automation so we could be cost effective while manufacturing in the United States. Stated another way, was it possible to substitute capital for labor and thereby be competitive with a shortened manufacturing cycle? It was a question no one had successfully answered in a positive way.

In February of 1985 we formally offered Peter the position of manager of production technology. In April he accepted. The plan was for him to move to the U.S. late in the year. With the demise of Vivitar, he needed to be in Taiwan to restructure the operation and put into place the organization to make glass elements only. This meant the 33,000 square-foot metal plant had to be sold and, we hoped, sold in a way that provided employment for the 200 people who worked in it.

Peter made it known we had a factory for sale. In a short time, interested buyers were looking at the facility. An American company based primarily in Japan called Corton Trading, thought it could make astronomical telescopes and electrical components there. One of its principals had a work history with a U.S. camera maker and had a number of U.S. optical design contacts. The firm at one time had a relationship with Vivitar. Its president was excited about the possibility of making all the metal parts and doing the assembly work on the 450mm MUM. Ironically, like Vivitar, Corton's interest in a Taiwan manufacturing facility was to free itself from being dependent on Japanese suppliers. In a brief negotiation, the metal factory was sold for $500,000. Because of worker transfers, social costs were kept to a minimum. It was a good deal for us because we couldn't use it, and it was a bargain for Corton if they could.

Representatives of Eastman Kodak heard that part of BASO was for sale, and a few weeks after the deal on the metal plant they visited Peter. Kodak had decided to transfer as much optics manufacturing as possible from Rochester, New York to the Far East. After looking at many optic manufacturers in the low-cost countries, the one they found most attractive was BASO. Peter informed us of the details of their visit and referred them to me.

In a Cincinnati meeting with Kodak executives, I explained that we had to have a low-cost source of glass lenses and that BASO fulfilled that need. They argued that if we sold them the glass factory they could use it for their requirements and also supply us, thereby making it even more cost-effective. Part of the deal would be for them to be a long-term source of supply at the same low prices we were enjoying. We told them we would think about it.

There were a lot of things to take into consideration if selling the BASO glass factory was a possibility. Would Kodak be a good supplier at reasonable prices? Our answer was, with carefully specified conditions—yes. Would our customers look at us differently because we didn't own a glass lensmaking operation? We didn't think it would be a problem if the arrangement was fully explained and we gave them assurances. Could we build a cost-effective glass optical factory in Cincinnati? Even with Peter's expertise, unless there was a breakthrough in processing technology, the answer was probably no. Could we continue to operate the glass lensmaking part of BASO profitably without the Vivitar volume that had been part of the original equation? There was not a clear answer to this one, but we thought the answer was yes.

Kodak kept pressing us to consider selling. We continued to analyze all the options. Given the logistical difficulties with Taiwan operations and having concluded that eventually we had to have a competitive U.S. capability, we entered into negotiations. After more meetings in October, we finally agreed to a deal. The price was slightly in excess of $4 million. We were assured continuing low prices and expanded capacity to serve future needs. The deal with Kodak, along with the Corton sale, resulted in an exceptional return on the $1.00 original investment. Nevertheless, we were not completely sure selling BASO was the right thing to do.

After the sale contract was signed, I asked a lot of questions about glass lens automation in discussions with Richard Kleinhans, the senior vice-president in charge of all Kodak manufacturing. From other Kodak executives, we knew their efforts in this area had been extensive. His answer to the possibility of automating large lensmaking in the U.S. was less than encouraging. He said, "Anybody who thinks it can be done is really stupid." Since that is exactly what we were thinking about, his words were very sobering. They made me wonder if such thinking was crazy.

For the balance of the year after the BASO sale, Peter wrapped up his affairs in Taiwan and prepared for the move to the United States. His first major Cincinnati activity was to purchase and install a lenscleaning machine. Two trips were made to the U.S. to get the project going prior to the move with his family. After being formed, large plastic lenses need to be cleaned, destatitized and dried before vacuum deposition anti-reflective coatings could be applied. This had to be done in a labor-intensive procedure that was costly and induced excessive scratching. Working with engineering associate Larry Gerding, several suppliers were evaluated and an order was placed for equipment late in the year.

Peter also did an extensive survey of lenscleaning and lensmaking equipment manufacturers in Japan. Nothing particularly exciting was seen. Two companies had heavily automated conventional grinding and polishing machines with simple pick and place robots. The systems were for lens diameters considerably smaller than those we were producing. He was not impressed with the equipment and did not think it could be modified to give us the technology advancement we were seeking. He also visited an Italian optical equipment company that made automated ophthalmic lens processing machinery. He found it to be too light in weight and limited in the range of curvatures it could process. Even if the machine could have been modified for our kind of curvatures, he was convinced the lack of sturdiness would have caused it to soon fall apart running large heavy lenses. All these queries failed to identify a suitable glass lens processing technique that would dramatically reduce labor costs and make it possible for us to efficiently produce glass lenses in the U.S.

Our extremely capable engineer, Jim Ballmer, was also investigating techniques that inventors in the U.S. were promoting. We concluded the untested ideas were simply too risky.

Over the year-end Christmas holidays, Peter and his family were in Austria. While there he talked to a friend who was the president of the LOH Company in Wetzlar, Germany. LOH had a reputation for building the finest lensmaking equipment in the world. It was also the most expensive. What he learned was extremely interesting. The LOH chief executive confidentially told him they had developed a completely new geometric technique for grinding, lapping and polishing lenses. The cycle time was very fast. It was treated as an important secret because LOH viewed the technology as having the potential to make European and American optics manufacturers competitive again. Very few people had been given the details about how the machines functioned.

Peter and his family arrived in Cincinnati in January and he immediately told me about the LOH development. We recognized it as a potential processing breakthrough that might make it possible for us to competitively fabricate lenses in Cincinnati. At the same time we were seeking a source to press thin aspheric glass lenses we called C-shells, which would be used as lens elements placed next to projection CRTs. The cavity between the C-shell and the CRT would be filled with a liquid for cooling and elimination of reflections. There also was a possible supplier in Wetzlar not far from the Frankfurt airport. Peter and I decided to visit both companies as quickly as possible. Within two weeks we departed for Germany.

The LOH Company was impressive. Their obviously brilliant engineers designed and built rugged equipment. The new technology was called Synchro Speed. In a major departure from conventional grinding, lapping and polishing techniques, it did not employ any oscillating movements. The lens blank was held on a lower spindle and rotated at a predetermined speed. A tool, larger than typically used in the old process, rotated at another precalculated speed. It had to be held at an exact angle. To make the process work, both the speeds of the element being processed and the tool had to be precisely synchronized—hence the name Synchro Speed. The pressure between the elements and the tool, as well as the speed of rotation, were much higher than the conventional process, resulting in a considerably shorter work cycle. A highly proprietary mathematical program to calculate all the parameters was treated as an important secret.

The process to make a lens required nine steps. A pressed glass blank, slightly oversize in curvature and diameter, is obtained from a glassmaker. Blanks have variegated surfaces far too rough to transmit light accurately. They are processed with hard abrasive or diamond tools. The steps begin with rough grinding of each surface. Then both sides receive a fine grind, sometimes referred to as lapping. Next, each surface is polished with a fine abrasive compound, followed by polishing with an extremely fine grain abrasive to achieve the final surface. The last step is to center the lens. It is performed on a machine that grinds the outer edge in a way that positions the peak curvature in the exact center.

Peter and I calculated the production throughput that could be achieved with the Synchro Speed system. It was impressive. A production line would require nine machines. The equipment was available in two sizes. The smaller 100mm machines processed lenses from 1-inch to 3.9-inches in diameter. The larger 175mm version could process 1.5-inch to 6.8-inch diameters. Although the machines had simple built-in pick and place robots for inserting and removing the blanks, each would require an operator. Processing, although much lower in cost with our labor rates than conventional lensmaking technology, was still more expensive than the Far East alternative.

We left Wetzlar, telling LOH we were quite interested but needed to study the matter further. On the way home Peter and I talked about all the ramifications of completely automating the LOH technology, not with the simple pick and place robots the Japanese and Italians were using, but with fully programmable robots. Although it would be far more complicated to use such automation, we would gain enormous flexibility.

Upon our return, Peter went about studying everything that would be required to install automated Synchro Speed lines. With a USPL-designed raw material feed system, programmable robots and a super structure to tie the system together, a complete line would cost about $1 million. With cycle times in the range of one to one and a half minutes for each step of the process, the production goal was 20,000 to 40,000 lenses per month the difference being whether it was done on a 100mm or 175mm line. Smaller sizes could be processed more quickly. Doing it would also mean we had to acquire a glass cleaning system and a vacuum deposition coater to apply anti-reflective thin films.

We recognized that a decision to go forward carried substantial risks. The LOH technology had no history of long hard operation in a production environment. A complicated process tooling capability would have to be learned. And, other than Peter, no one else at USPL had glass lensmaking production experience. Those were the concerns. On the plus side we had concluded that when everything was taken into consideration, such as shipping, high inventories and labor, fully automated Synchro Speed lens production would provide us with lower-cost lenses than could be purchased from BASO—that is, of course, if we could make it work. The need for high volume Cincinnati glass lens production capability was so great we decided to take the chance. Everything that was needed to build one 100mm line and one 175mm line was ordered from LOH for delivery in 1986. The plan was to produce three or four long-running items that required few line changeovers and buy the balance of requirements from BASO. The lines would be installed in the new McMann East factory.

Simultaneously, Peter put together a team to do all the things necessary to get us into production once the equipment arrived. It included Mike Scarpa, a talented young engineer who worked on automation. It was determined that a central conveyor through the center of the line would feed to and receive elements at every transfer point from a series of Hitachi robots. Mike Reinert, an experienced lensmaker, was hired from Perkin Elmer Corporation to work on the manufacturing process. Randy Jennings led the group that took on the responsibility for installation and maintenance to keep it running. Jill McClure and Serpil (Sam) Dayi were trained to make the specialized tooling necessary to support production. With the help of Larry Gerding, Peter determined the specifications for the lenscleaning system, selected a maker and placed the order. He also investigated vacuum deposition coater manufacturers and ordered a specialized machine costing several hundred thousand dollars for that task. In total, commitments were made to spend upwards of $3 million.

The move into glass lensmaking in Cincinnati was a major undertaking. We knew it would be difficult. Later we would learn just how difficult it would be.

# 56 1985

The final realization that there would be no future with Vivitar, and what it meant to the Taiwan operations, had led to the sale of the BASO metal plant to Corton. The logistical problems of obtaining all-glass elements from Taiwan led to the sale of the glass factory to Kodak, and once again motivated us to find a way of efficiently making high volume glass lenses in Cincinnati. These events, which took place in just a little over one year, required major shifts in strategy and business plans. We ended the year with a very different view of how the company would be configured than we had held at its beginning.

Our obsession with the undervalued Japanese yen and the competitive threat it posed for USPL, was reflected in front page *FOCUS* articles that I wrote in March and Dave Hinchman wrote in December. Four Japanese TV lens and two standard optics competitors were making offerings in the market. The patent encroachment activity was another challenge that overlaid the currency problems. For the year, the yen fluctuated between 200 and 263 against the dollar. USPL prices varied 31% for our Japanese customers because they were paying in dollars. When it was in the 260 range, we heard a great deal about potential Japanese lens competitors.

Fiscal 1985, ending June 30, was both a financially excellent and frustrating year. Sales were $31.9 million—up 54%. Although we were a Sub Chapter S corporation, with profits flowing through individual shareholder tax returns, for comparison purposes they were always viewed on a net after federal and state tax basis. Profits, excluding a special item, were $3.4 million. The tax act of 1984 eliminated domestic international sales corporations, commonly called DISCs, that allowed certain taxes on export profits to be deferred. The legislation made it possible to take into income in 1985 everything that had been deferred through 1984. For USPL this amounted to $660,000.

That amount, added to the $3.4 million, brought the profit to over $4 million—an increase of 42%.

Standard optic volume was down slightly for the year. However, a very high level of development activity and prototyping for fiber optic connecting lenses caused Stretch Hoff and his associates in that division to believe it would become a big business. Hewlett-Packard was the largest customer. Much promising work was going on with ADC and AMP for similar lenses. Standard optics again ratcheted up production of molded housings for the huge jump in television lenses. PTV sales increased from $17 million to $29 million—up 70%.

It had not been an easy year. The large sales increase, coupled with the introduction of lower-priced new lens designs such as Compact Delta 7, meant throughput of assemblies was higher than the percentage sales increase indicated. All of this resulted in excessive overtime, poor yields and too much scrap. Top managers were distracted with everything that was going on at BASO and the planning for the new McMann East factory. For about nine months everyone referred to the struggle as the "frenzy." Finally, near the end of the fiscal year, in April the problems were eliminated and the "frenzy" ended. Our technical and manufacturing associates at all levels were somewhat exhausted by it all, but they bounced back quickly.

At the height of the "frenzy" we had 845 associates. Over 200 were temporary. When the problems subsided, the total dropped to below 700. We had long appreciated the importance of persuading customers to place orders with delivery releases that allowed us to schedule production in an orderly way. It was becoming even more important because a highly fluctuating workforce caused too much variation in skill level. PTV makers' sales to consumers were seasonal and consistently peaked in the last part of the year. In 1985 40% of the year's volume was sold in the last quarter. We had to continually push the makers to build inventory in the early part of the year so that we could level load our factories.

For the calendar year, PTV sales by manufacturers to retailers were up 36% to 266,000 units. In "Projection Television Past and Future," we reported that near the end of the year some leading brands were sold out. We further reported with notable exceptions, that manufacturers were inconsistent in their sponsorship of PTV and failed to aggressively promote the category. The jump in volume had come with little advertising and promotion in the media. Ninety percent of the year's sales were rear-screen projectors, with the other 10% being larger front-screen two-piece sets that were ceiling mounted or built into coffee tables.

In late April we had a meeting at the Coldstream Country Club attended by 110 managers, engineers, technicians and supervisors. I reviewed what we believed was a very positive future outlook. Having just come through the "frenzy," everyone needed to hear good news. The primary thrust of the session was to talk about our competitive position, and what USPL would have to do if we wanted to remain the dominant maker of PTV lenses. Having worked closely with the Japanese for so many years and believing they would be our only viable competition, I talked extensively about how they would come after us. To every extent possible, they would copy our designs. They would engineer products in a few narrow areas that would be high in quality and low in cost. In short, they would not carry our burden of having to supply a wide range of designs but rather "cherry pick" a couple of big volume items. Once they were in production there would be aggressive cost reduction efforts through better engineering, attention to detail and quality improvement.

To remain number one, I told everyone we would have to do all the things the Japanese competitors will do plus continuing to provide the wide product line. It meant continual improvement in every facet of our design, engineering and manufacturing efforts. In virtually every Japanese factory the goals for the year are summed up in a slogan. We would do the same thing with a program called "Staying

#1." USPL had always striven for continual improvement. The program that was introduced at this meeting was different in that it specifically formalized a process to chart and display all critical measures of performance against goals. It was to graphically show associates in each operation how they were doing in relation to the past and against an improvement objective.

Dr. Akio Ohkoshi, the revered Sony senior scientist who was a principal inventor of the Trinitron picture tube, came to Orlando, Florida in June to receive a prestigious technical award from the Society of Information Display. Fred Ishii, a young Japanese San Diego-based executive who spoke fluent English, was given the assignment to assist him in his travels. Dr. Ohkoshi had always been supportive of USPL since our first trip to the Tokyo headquarters and over the years he had regularly entertained us. Brian Welham and I saw this as a chance to repay the favor. We invited Dr. Ohkoshi and Fred Ishii to my home in Boca Grande, Florida on the Gulf of Mexico for a weekend of relaxing and tarpon fishing. They accepted. Brian flew with them from Orlando to Sarasota where we all met for the one hour drive south.

Dr. Ohkoshi needed a bathing suit, so we made a stop to find one. Because he was such a small man, procuring the right size was a problem. Eventually we found a boy's suit that fit. From there it was on to get groceries. I asked Dr. Ohkoshi if he would like to see an American supermarket. He was eager to do so. After being in the store about five minutes, I noticed the other shoppers had stopped and were staring at us. Looking back I saw why. Fred Ishii had a large video camera on his shoulder and was filming Dr. Ohkoshi and the market. He quickly explained to me that one of his responsibilities was to film Dr. Ohkoshi's trip to America. A moment later the store manager rushed up to ask what was going on. I said this is Dr. Ohkoshi, the prominent Japanese inventor of the Trinitron picture tube, and the other fellow is filming his first visit to a U.S. supermarket.

The manager was effusive in his welcome to the man whom I had just billed as a celebrity and then quickly disappeared. Moments later he returned to personally conduct a store tour. When we reached a beautiful display of cheeses, a gracious clerk asked Dr. Ohkoshi, by name, if he would like to try any of the items. From there it was on to pastries where another clerk, again calling him by name, offered samples. The tour continued with the manager doing everything possible to make the experience enjoyable for Dr. Ohkoshi. When we left the store I explained to him that this was not a typical visit to a U.S. supermarket, and that I had never seen anyone treated so graciously in such an establishment. He thought the experience was wonderful.

At quiet Boca Grande we boated to nearby islands and swam in the ocean. Two evenings were spent tarpon fishing from 8 o'clock until midnight. Nothing had been caught until the last night with about fifteen minutes to fish. Dr. Ohkoshi hooked a huge tarpon that clearly weighed more than he did. It nearly pulled him out of the boat. I had a fleeting vision that I would have to call Sony Chairman, Akio Morita, and say something like, "We did everything we could to save him." He fought the fish for nearly forty minutes, finally bringing it to the side of the boat. It had been an incredible tug of war with the lighter opponent winning. After fearing we would have no luck, it was a great end to the trip. He couldn't have been more delighted and, as instructed, Fred Ishii had recorded it all. A couple of weeks later I received a warm thank-you letter from Dr. Ohkoshi in which he said, "Through my life, I never had such an enjoyable trip." He would often repeat those words in the years to come.

Economic conditions in 1985 were generally a little better. The bank prime lending rate dropped about one and a half percentage points. It ranged from 9.5% to 10.6%. Inflation decreased from 4.3% to 3.6%, while unemployment remained flat to down slightly at 7.2%. The personal savings rate fell from 8.6% to 6.9%. Overall, the future looked good.

# 57 A Pivotal Review and Look at the Future

At the beginning of 1986, we had an unwavering conviction that there was a growing desire by consumers to view TV images on large screens. There was no doubt in our minds that projection television would continue to grow into a significantly bigger business. It was the overriding positive factor about USPL as we looked to the future. There were, however, concerns that carried negative connotations. Television industry, competitive and economic risks had to be reconciled as we planned for the days ahead. And, I quietly had begun to think about how, as well as when, USPL ownership might be passed on. The Howe family held a large majority of the shares. It was an estate planning issue I knew would eventually have to be resolved in a way that was good for the people in the company, our customers and the family.

Customer concentration was an ever present concern. Nearly 80% of the company's total business was with five companies. Fifty-two percent of it was with companies in Japan, where TV lens competition was emerging. USPL market share dominance was being targeted by glass lensmaker Norita and plastic lensmakers Hitachi, Fujinon and Matsushita. Reports circulated widely that a number of other Japanese companies were interested in the projection TV market. Japanese lens designers who had markedly increased their skills carefully studied the Opcon designs and worked to emulate USPL optical performance while attempting to circumvent our patents. They had not yet succeeded, but substantial efforts were bringing them uncomfortably close. It was clear there would be intellectual property challenges, and they would be from giant companies that had unlimited resources.

The notoriously competitive conventional television receiver market was wonderful for consumers who were able to buy better and considerably larger screen size sets at lower prices than had been paid

twenty years earlier. And, this was after a period of extremely high inflation. For TV manufacturers, the cutthroat competition had a very dark
side. Virtually no TV maker earned a reasonable profit, and most were
losing large sums of money. They depended on VCRs, video cameras
and PTVs for better profit contribution. However, the margin levels for
those products were also pressed. USPL customers' lack of profitability
was troubling because it caused an uncertainty for the continuing via-
bility of some of them. We had already seen Motorola, Westinghouse
and Admiral exit the business, along with others. The fact that we were
quite profitable and our customers were not was a principal factor in
dissuading us from considering a public offering that would provide an
option for an orderly change of ownership. Revealing our profitability
would have led to greatly increased downward pressure on lens prices.

Finally, there was the dollar-yen currency fluctuation issue
that had been with us for so long. We were quite aware that a grossly
under-valued yen could cause havoc in our business by making it possible for inferior manufacturing technologies to have competitive advantage from a cost standpoint. Again, this was a huge concern.

These were the big issues that had to be grappled with as we
looked ahead.

For its February 3 program, the Cincinnati Young Presidents'
Organization brought in a speaker by the name of Ram Charan, who
had a superb reputation as a consultant on international competitive
strategy. Among his clients were IBM, Hewlett-Packard, Motorola
and Corning Glass Works. He had a fascinating way of asking questions about one's business and competition that would lead to his re-
casting the questions in unique ways, thus causing people to broaden
their thinking. I was very impressed with his presentation because
many of the strategic and competitive things he talked about were
issues that deeply concerned us. We had already planned an offsite
management retreat for April on the same subject in Boca Grande,
Florida. I asked Charan if he would meet with us for a day to take our

group through the analysis that had been presented to Y.P.O. but in a way very specific to USPL. His price was, at the time, a choking $8,000, but I thought it would be money well spent. We agreed that I would send him financial, market position and product information before the meeting.

The four-day meeting began on a Friday. The plan was to meet that afternoon, all day Saturday with Charan, Sunday morning, and finally again on Monday morning before returning to Cincinnati in the afternoon. Our ten top operations people were Ron Antos, Jerry Behne, Jack Collins, Pete Cutler, Dave Hinchman, Stretch Hoff, Len Kosharek, Peter Scherb, Brian Welham and me. All were in attendance. Except for the day with Charan, our sessions involved capacity, personnel, financial and organizational issues. From time to time we found it useful to get away from the factories and phones for such planning meetings involving these and other subjects.

Ram Charan, who I believe is a native of India, is very interesting. He is the only person I have ever met who literally has no home, and lives throughout the year out of a suitcase. He said that since he was not married, and was constantly traveling from client to client, it made no sense for him to have a home. It was hard to believe anyone could live that way. He did have a secretary in Dallas where clients could leave messages. The only reason we were able to get a time on his extremely full schedule with such short notice was because we were meeting over a weekend.

Charan asked voluminous questions about the TV market, our customers, competition and technology threats. With those answers outlined on easel pads, he asked related questions involving who we thought the dominant companies would be in five years, how they might look at vertical integration, and how the technology might change. It was all very interesting as we were forced to conclude who the winners and losers might be and what the various scenarios would mean to USPL. Japan's huge Matsushita, the world's largest consumer electronics manufacturer, was a particular concern because of its lens-making efforts. We knew that if they decided to make an aggressive move into lens manufacturing, they would be able to pour virtually unlimited resources into its execution. Hitachi also caused us worry for the same reasons. The poor financial health of Zenith and North American Philips was also troubling. RCA had recently been acquired by General Electric, bringing uncertainty to its future. G.E. had notoriously high margin standards, but it had been many years since they made a profit with television. RCA had the largest U.S. market share, but also suffered with lackluster profitability. It was difficult to envision what G.E. could do to fix the television part of its business. The conclusion was that they made the acquisition to get RCA's medical imaging and defense electronics divisions, and they had to take the TV business as part of the deal. At this point, it looked to us like there was a possibility the television receiver industry could become totally dominated by Japanese companies. Every one of their major manufacturers had assembly operations in the United States or Mexico or both. Matsushita, Sony and Toshiba had made capital-intensive picture tube manufacturing investments in the U.S. that were state of the art in design.

The Charan exercise didn't really tell us anything we hadn't already known. It did cause us to look at the risks and opportunities from different perspectives and in that sense, was useful. As we had before, we saw the customer concentration, currency fluctuation and Japanese self-manufacturing trends as serious issues. Charan also recognized that we had a big lead with unique manufacturing technology and design patents. The conclusion was that these strengths would not stop the Japanese giants from coming after our business if that is what they chose to do. It was a sobering day for us as we looked into the future trying to predict what was going to come.

Near the end of the day, Charan took Dave and me aside for a private conversation. He said we had built a wonderful company that seemed to him to have an excellent future. However, the mammoth companies in the industry we served had a history of vertical integration. Thus, with our lack of diversification, the risks were substantial. He added that he knew of a company trying to broaden its business which would be a perfect fit with USPL. It was Corning Glass Works of Corning, New York—a company that had a large television presence as a supplier of CRT funnels and faceplates to tubemakers. He suggested that we talk with them and offered to call a contact there to recommend they talk with us.

I found it ironic that he thought USPL would be a great acquisition by Corning. In the early 1970s, a friend and I discussed who we thought might be logical strategic acquirers for USPL someday, and I said I could only think of two companies—Corning Glass and instrument-optic maker Perkin Elmer. Now Charan had made the same conclusion about one of them.

Our reaction to his suggestion was that the idea was worth preliminary exploration if there was a genuine interest on Corning's part and if there were compelling strategic reasons for both companies. He placed the call to a Corning executive he knew the following week to tell him what an exceptional company USPL was and why he thought they should talk to us.

# 58 Fiscal Year 1986

The July 1, 1985–June 30, 1986 fiscal year was similar to the strong previous year. Sales were up 5% to $33,615,000, and net after-tax profit, before special items, was up 2% to $3,481,000. The after-tax gain on the sale of BASO was $2,649,000, which gave us a total profit of $6,130,000, up from the FY '85 all-inclusive profit of $4,071,000. It was a gain of 50%. Dividends, in addition to Sub Chapter S tax distributions of $3 million, were $5 million. This left us with year-end cash of $8 million on the balance sheet. Capital expenditures had dropped to $2 million from the previous year's $4,400,000.

The yen had continued to gain in value against our currency. By April it was 180 to the dollar, making our lenses at least 10% less expensive for the Japanese TV companies. It had been a 30% favorable change for us from the previous summer. Federal Reserve Bank Chairman, Paul Volcker, and leading economists were saying the yen should gain more. The consensus was the currencies would reach equilibrium at 160 yen to the dollar. We were surprised that the pressure on us to reduce prices did not subside more than it did. We were also surprised by the apparent ease with which Japanese industry adjusted to what one thought would have been a more difficult competitive environment. It caused me to wonder if they had been far more profitable than financial reporting indicated as a result of differences in accounting methods. In any event, the change was good for our company.

In March, USPL was given a significant recognition by Sony. It was their first U.S. Award of Excellence. We were one of ten recipients and, importantly, the only one that shipped a high volume of products to Japan. Engineering and procurement managers in San Diego and Tokyo had nominated us. An elaborate award ceremony took place in San Diego. Special messages of congratulations were received from executives in Tokyo who wanted it made clear that, in their view, we were uniquely special because the award was for superb performance in both the United States and Japan. To be chosen, a company had to score above 90 in an elaborate grading procedure measuring quality, delivery reliability, competitiveness, engineering, production management and quality control systems. The effusive recognition, by such a prestigious customer as Sony, was gratifying to all associates.

That evening I met co-founder Akio Morita for the first and only time. I knew his brother and many other top-level Sony executives, but not this world-renowned visionary. When learning I was with USPL, he graciously expressed appreciation for our role in supporting Sony. The knowledge he demonstrated about the company made it clear he had heard much about us over the years.

In 1986, all USPL associates participated in what was called a climate survey. An outside personnel consultant with expertise in conducting and tabulating such surveys, was employed for the task. The objective was to learn, on an anonymous basis, how our associates felt about USPL as a place to work, training, internal communications, company policies, supervision consistency, how they perceived they were valued as individuals, benefits and compensation. The only answer identification related to the department in which the respondent worked. The solicitation of this kind of feedback from employees is quite common now, but it was not then.

We found the exercise to be quite useful. In general, the results were gratifying. Most associates thought USPL was an excellent place to work, with good pay, good benefits and fair policies. There were, however, some negative comments that could be traced to certain departments where supervision was not as respectful or consistent as was dictated by our policy. I reported the results to all associates at company meetings in an unvarnished way. They were frankly told what had been learned—both good and bad. To be completely sure of what was being heard from our associates, we employed a couple of personnel experts who spent two weeks making in-depth visits to every department on every shift. Everyone had a chance to further tell us what they liked and didn't like about the company in an anonymous way. The feedback from the survey and the experts was consistent. The problems uncovered were just in a few areas, and management soon corrected them.

USPL has continued to do climate surveys over the years at appropriate intervals. They have been extremely helpful as another way to positively improve communication with associates.

Projection television receivers had been sold to the public for over ten years, yet little market research had been done on consumer attitudes about the product. In every customer meeting we would talk about market trends and exchange thoughts about the category. Unfortunately it was all subjective. Jack Gifford, the outstanding Miami University marketing professor who had held a class study on whether we should enter the hand-held magnifier market years earlier and I discussed the idea of him having a class do a PTV research project for which USPL would foot the cost. He was intrigued by the learning experience it might provide for students and after a few weeks of thinking about it, enthusiastically embraced the idea. In the spring, forty-two students registered for two marketing classes that would perform the study. The plan was to do a national consumer survey, field experiments in shopping malls utilizing projectors to test consumer attitudes, and conduct focus group interviews with potential buyers. Our customers were very enthusiastic about the project and agreed to cooperate by providing projections sets that could be used in the field experiments and address lists of previous PTV buyers for follow-up interviews on their experiences with the product.

The class's primary objective was to determine what activities should be undertaken by manufacturers, wholesalers and retailers to increase the profitable sales of rear projection television to consumers over the coming three years. Using a *TV Guide* subscriber list, 2,560 people were randomly surveyed. Two focus group interviews were conducted in a commercial market research laboratory. In the field study, 275 people were surveyed at shopping malls after viewing PTVs. Fifteen students held meetings in our Ambassador East Hotel suite during the Consumer Electronics Show in Chicago with top marketing executives from RCA, Philips, Panasonic and Zenith to get their input. Twenty-four retailers and wholesalers were surveyed, and finally 300 PTV owners were called to learn their level of satisfaction with what they had purchased.

The students' concluding recommendations were backed up with massive amounts of data. They said there was a lack of consumer awareness about the product and some negative impressions existed because of early comparatively inferior sets that had been sold. They advised manufacturers to heavily advertise the high quality of what was currently being sold in simple terms. Further, they said it should be promoted as "big screen" rather than "projection." They determined that retail sales people needed better training, and that store displays optimizing viewing demonstrations should be improved. Proper set adjustment in stores was an obvious problem that required attention.

Although the manufacturers were well aware of the product marketing problems, the substantial data generated by the students caused them to reassess their advertising efforts. We don't know how much of it was related to the study, but in the years to come budgets were significantly increased. We and our customers thought the students had done a very professional job. USPL received many accolades for sponsoring the study which represented yet another way our company provided value to the people with whom we did business.

# 59 Corning

Ram Charan's call to Norman Garrity, Corning's Lighting Products Manager, brought an immediate response. Within a few days we were to be visited by David Spille, a business development executive, whose mission would be to learn what he could about USPL's technology and market position.

Corning Glass Works, founded in 1851, was a technology and consumer products company widely admired by both industry and investors throughout the world. We had previous contact with them as a result of a brief visit during the search for glass blank suppliers, and also during a Cincinnati meeting arranged by Opcon where they informed us about a precision glass molding technology being perfected to make small aspheres. Corning had a superb reputation for research and innovation. We were aware of the fact it was the company that perfected the process to manufacture ultra-pure glass, making possible the development of high performance fiber optics which were revolutionizing the telecommunications industry. A single optical fiber could carry 200 times as much information as a pair of copper wires at that point in time. They had developed the first CRT glass faceplates and funnel components for the television industry, invented photochromic ophthalmic lenses that darkened when exposed to sunlight as well as virtually unbreakable ceramic tableware that was sold under their own brand name.

Prior to Spille's visit, I obtained an investment analyst's report and a copy of the previous year's annual report to find out more about them. They had a large medical laboratory blood testing business, made lighting glass for a multitude of applications, ceramic substrates for emission control systems, glass laboratory products and owned Steuben Glass. They were also 50% owners of two highly successful joint ventures—Dow Corning, a company renowned for silicon technology, and Samsung Corning, a maker of TV CRT glass products in Korea.

Investor interest was centered on their fast-growing fiber optics business, which brought substantial gains in the value of the company's New York Stock Exchange traded shares. Sales for the year were projected to be $2 billion, and they were expected to earn $180 million. Corning employed 25,000 people.

Just the thought of selling the company into which we had invested so much of our lives in building, brought mixed emotions to the surface. The business was doing well, and the future outlook was excellent. If selling was something that was seriously going to be considered, there would have to be unquestioned cultural compatibility and meaningful tangible benefits that each company brought to the other. And the price would have to be substantial, reflecting the growth potential of USPL. These things that were going through my mind before Dave Spille arrived on April 11.

In this initial meeting we gave him a general overview of our technology, market position and customer makeup. He was taken on a tour similar to those we gave to visiting customers. Areas of particular common interest were fiber optic connectors, glass lens blanks, PTV CRT faceplates, auto-focus camera optics and automotive lighting. Dave represented Corning well and left us with a good impression.

Two weeks later he returned with Bob Ecklin, vice-president of business development, and four other people with manufacturing, research and financial responsibilities. Bob was the leader of the delegation whose objective was to probe deeper into USPL operations and to give us the opportunity to probe deeper into Corning as we looked for "compelling reasons" for taking discussions to another level. He told us they commissioned a strategic materials study by the Arthur D. Little consulting firm that had concluded Corning should be far more involved with plastics—particularly plastics used in optics. USPL had already been identified as a company with such expertise. Our fiber optic connector lens business with Hewlett-Packard and ADC, plus development programs with AMP, ITT and GTE, were particularly interesting to them. Those customers were forecasting huge demand for such plastic lenses. Making the PTV CRT faceplate an integral part of the projection lens optical design was another area of common interest. The automated glass lens fabrication lines on order from LOH would have to be fed with lens blanks. There was a marked change from the apparent lack of interest in supplying such optical glass from our first visit years earlier. They thought this was an area of mutual opportunity to gain competitive advantage.

Bob Ecklin was bright, easy to communicate with and openly frank about Corning's growing interest in USPL. The executives he brought with him were also impressive. He indicated they would like to have us meet other Corning people in a meeting that would be arranged soon.

My schedule was so fully packed we could not get together again until May 20. Ecklin and Spille, accompanied by four more people who were specialists in the areas of common interest, came for technology discussions. By this time we were comfortably exchanging a great amount of information, including financial data. Their interest and ours was growing, but we were a long way from specifically discussing price. I had told Bob that because of our growth record and future projections, we should not continue discussions unless they recognized that our company had very substantial value. He concurred and agreed that it did.

At this point my view was that the discussions had a good chance of being no more than simply an interesting exercise, and if nothing else, might lead to our finding out USPL's value to a strategic buyer. The previous year at a mergers and acquisitions seminar, I had met an impressive young man who was a senior executive with a New York investment group that negotiated conservatively leveraged buyouts of companies. Although we had no interest in selling the company, we engaged him to tell us what the company might be worth to a financial buyer. His detailed report stated the value was in the range of $37 million, assuming a 10% annual growth rate. This valuation assumed excess cash had been dividended out to shareholders. We concluded that given general valuations at the time, his numbers were probably in the right area for a financial buyer who would look at the company on a stand alone basis. For a strategic buyer who could see inter-relationship benefits accruing to its business, I believed USPL was worth in excess of $50 million. As the discussions were proceeding, these year-old benchmarks were in the back of my mind.

Remembering how I was naively negatively affected when Scott Paper acquired my first employer, S. D. Warren, I was particularly sensitive to how our key executives might feel about a merger. If a deal had a good chance of becoming a reality, I had vowed they would be part of the decision-making process. With a fourth Corning meeting about to take place, it was time to tell them what was going on. I swore the managers to secrecy and confided the details of all the discussions to date. There were ten of us who knew what was happening; the plan would be to keep them completely informed as talks moved forward. I told them that even if this was something I decided was right for the company, we would not do it if they were not in agreement. They would have the right to veto a deal. This made it possible for all of them to be involved in the discussions, the benefit being that Corning people could have frank technical dialogue with them and make judgments about what I believed was our very impressive leadership group. It also allowed our executives to make judgments about the Corning people as well as the pros and cons for a joining of the companies.

On June 18 and 19, I was invited to Corning to meet with their senior executives. Jack Collins accompanied me. He could provide in-depth financial data, and I was interested in his observations about the sessions and the people with whom we met.

The leader of the group who received us was Roger Ackerman, group president in charge of specialty glass and ceramics. Roger was in charge of one of the four operating groups reporting to Chairman and CEO, James Houghton. Others present were Van Campbell, chief financial officer; James Flynn, treasurer; Ken Freeman, controller; James Riesbeck, head of strategic planning; William Ughetta, general counsel; and Bob Ecklin. I was asked to bring along a variety of samples that represented the kinds of products USPL manufactured. After a get-acquainted meeting in which the samples were displayed, dinner was served during which frank questions were asked by both groups about each other's businesses.

The next day Jack exchanged information with financial people, and I spent the morning getting to know Roger Ackerman. A long luncheon meeting with Jamie Houghton, which included Roger and others, took the discussions to a new level. Jamie was knowledgeable about USPL and, along with Roger, made it clear they were interested in acquiring our company if a deal satisfying everyone's needs could be negotiated. Jamie and Roger impressed me as pleasant, straightforward and clear-thinking as to where they wanted to take Corning. In an internal booklet Bob Ecklin had given me, I had been pleased to see a strong statement delineating corporate purpose, quality objectives, uncompromising integrity standards, technology objectives and the importance of valuing individuals who made up the workforce. It was in many ways similar to the Business Philosophy of U.S. Precision Lens. Jamie Houghton, impressed with our values statement, pointed out the similarity to a number of his associates.

As the visit to Corning ended, it was agreed we would meet again in Cincinnati a week later, on June 24, after they had a chance to analyze the extensive USPL financial data Jack had provided.

On our return, the USPL board and key managers were briefed in detail about every aspect of the discussions. There was unanimous belief on the part of the board that we should seriously consider being acquired if the price was attractive. The key managers, in general, were positive but less enthusiastic than the board because they would have to work with Corning people they did not yet know much about. It was understandable apprehension. I did my best to tell them everything I knew and didn't know. We were involved in a process that did not yet require any firm decisions to be made.

There was a reason to move the negotiations along rapidly. Congress had passed a capital gains tax increase that would raise the rate from 20% to 28% effective January 1, 1987. Since one of the principal motivations was asset diversification, we had already determined that if a deal was consummated involving an exchange of stock we would sell most of Corning shares. To do it before the new rate was effective would save us nearly 8% of sale proceeds.

The June 24 meeting was held in downtown Cincinnati, at the offices of our accounting firm, Arthur Andersen. Representing USPL were Dave Hinchman, Jack Collins, our attorney, Don Lerner, Dave Phillips, the managing partner of Arthur Andersen in Cincinnati, and me. I recognized Phillips as a skilled negotiator as well as a brilliant accountant who could be particularly helpful in structuring an agreement in a way that would maximize the tax advantages for both our shareholders and Corning. I had complete confidence in Don Lerner, and had no doubt he and Phillips would make an excellent team. Corning was represented by Roger Ackerman, who would be responsible for USPL if it was acquired, Bob Ecklin, Treasurer Jim Flynn, Controller Ken Freeman, Strategic Planning Manager Jim Riesbeck, and General Counsel Bill Ughetta.

Flynn spoke for Corning. He explained how they used different valuation techniques and elaborated on how risky USPL was because of its narrow product lines as well as customer concentration. He said their methods valued the company from a low of $45 million to a high in the range of $55 million. Being surprised, and somewhat irritated, I said something to the effect that, "if these were the kind of numbers we would be talking about, there was no reason to continue the meeting." Ackerman wisely suggested he and I leave the session for awhile to let the others continue talking. He and I visited an old friend of his who managed the Brooks Brothers clothing store next to Andersen's offices.

We returned to the meeting about an hour later to find the discussion centered on deal structure issues. Corning didn't want to do a cash transaction because they would have to put substantial good will on their balance sheet requiring an annual charge to earnings as it was amortized. They preferred to pay in stock, using an accounting method called pooling of interests. This avoided the annual charge to earnings. We did not want stock because if it was held past the first of the year the higher capital gains tax rate would have to be paid. Phillips proposed a method none of us ever heard of called a taxable pooling. A taxable pooling required that 5% of the price be paid in cash, with the balance being paid in common shares of stock. Such a technique would trigger the capital gains taxes for our shareholders at the time the deal closed, assuring the lower rate. Corning would be asked to give our shareholders who took stock a Securities and Exchange Commission registration as quickly as possible so shares could be sold. The required waiting period would be forty-five days from the close. This structure allowed Corning to avoid the balance sheet goodwill and amortization. Phillips and Lerner also forcefully argued that our patents could be written up in value in a procedure that would provide significant future tax savings for Corning. This was all very interesting but I thought probably academic, since there was going to be no transaction unless they had a major shift in thinking about price. We parted with the understanding they would review their valuation in light of what was learned in the session.

The next meeting was held in New York on July 2, during which Dave, Brian, Jack and I would hear their new offer if there was to be a new offer. On that day Corning stock was selling for $73 per share. The Dow Jones Industrial Average was at 1909. Roger Ackerman told us their offer was 993,000 shares less the 5% that would be paid in cash. This calculated to $72.5 million. If Corning stock fell in value within limits, additional shares would be issued. The conditions required approval by both boards of directors and a satisfactory due diligence investigation that would verify USPL was as it had been portrayed to them.

Given the conservative way businesses were valued in those days, I thought it was an excellent offer. We had just distributed $8 million in dividends to our shareholders, $5 million of which exceeded their tax obligations. This left us with a book value of $15.5 million, $8 million of that was in cash. At a price of $72.5 million, the transaction would be 21 times the previous year's earnings, 4.7 times book value with the cash, and 9.7 times book value without the cash. The high price meant the Corning executives believed our growth projections for PTV and capability to remain the dominant maker of lenses that would be required for that market were correct. The other products, particularly fiber optic connectors, were interesting to them, but the valuation was entirely driven by the prospects for projection television. Subject to the conditions of the offer, we had an agreement. Due diligence would start immediately, and details of the purchase agreement would be negotiated for a signing scheduled for August 15 and a deal closing on September 4 in New York.

On July 5 and 6, Jim Riesbeck, Bob Ecklin and I went to Chicago to visit with NEC, Matsushita and Zenith high-level executives as part of the due diligence. They wanted firsthand knowledge about the quality of USPL's customer relationships. A deal would not be announced until the purchase agreement was signed, so all I told the customers was that we were working closely with Corning on glass blank supply and CRT faceplate developments that, hopefully, would benefit our PTV customers. I explained that these fellows wished to understand the PTV business and the USPL position in it in greater depth. The customers were enthusiastic about the business and universally flattering with their comments about USPL.

Roger Ackerman knew that in the final analysis I would not do the deal unless there was agreement by the other nine top people at USPL who were privy to what was unfolding. Likewise, he did not want to make the acquisition unless he could have a high confidence level those key executives would remain with the company. On July 9 we had a dinner meeting at the Coldstream Country Club where he told the ten of us about Corning, its management philosophy, how he managed and why he thought a merger would be good for both companies. Our people's questions were frank and intensive. He made an excellent impression, causing the doubters to become quite comfortable with the idea of Corning becoming USPL's parent company. If this extremely important meeting had not gone well, negotiations would have ended. It gave Roger a chance to get to know people who were vital to USPL's future, and in doing so he gained a higher level of comfort.

On July 15 and 16, Jack Collins, Don Lerner and I went to Corning to work on the purchase agreement. On July 29 and 30, I was in New York with Brown Brothers Harriman, Goldman Sachs, Lazard Freres and Merrill Lynch investment bankers to learn what the options were to optimally sell such a large block of stock. While all this was taking place the overall stock market, and Corning stock in particular, was falling in value. Corning's price was highly influenced by the state of its fiber optic business. While the future outlook for the use of fiber was superb, at that point in time demand had diminished somewhat from previous months, causing investors to question Corning's valuation. Although the agreement called for them to issue us more shares if the price fell, there was a limit and it had been reached.

On August 6 I traveled to our place in Florida for a four-day stay to think about whether we should proceed. That day Corning stock closed at $59.83, down 18% from the $73 it had been on July 2. In the same period the Dow Jones Industrial Average had dropped 6.8% to 1779. At $59.83, Corning would issue 1,044,710 shares with a value of $62,500,000—a reduction of $10 million in just a month. Bob Ecklin and I had been talking a few times a day about this issue and other problems. I told him the deal was in jeopardy and we should meet. He flew to Florida for discussions. Correctly anticipating that we would break off negotiations, he received authority to raise the offer 11.7% to 1,166,810 shares for a value of slightly under $70 million. Roger Ackerman added a provision that put an additional $1 million at my disposal for distribution to our shareholders as I saw fit at the end of five years. Bob's visit kept the deal alive.

On August 15 Roger and I signed the purchase agreement that called for USPL to become a wholly owned subsidiary of Corning. Being a subsidiary corporation with a board process was important to me because I had seen how the division status of S. D. Warren blurred the lines of accountability and led to unwanted incursion by staff people in the name of help when it was acquired by Scott Paper. USPL was operating smoothly, and our managers didn't need to have a lot of their time being consumed by outside staff. Roger Ackerman agreed. He said our job was to continue to operate USPL as we had, and his job was to keep the Corning staff away from us. Another good decision by Corning was to continue using the U.S. Precision Lens name. Even though I always felt it was long and perhaps a little presumptuous, it was well known to the people in the markets we served, and therefore was of considerable value. On the day of the signing, Corning was selling for $60.75 a share for a deal value of $70,884,000. The Dow Jones Industrial Average closed at 1855.60, and the Standard & Poor's 500 Stock Index closed at 246.25. Press releases announcing the combination were issued by both companies. The price was not revealed.

On Monday, August 18, I told all of our people about the sale in a series of meetings. The formal presentation was as follows. After it I answered their questions.

### R.L.H. EMPLOYEE TALKS AUGUST 18, 1986

The purpose of this meeting is to make a major announcement that we believe is very positive for the future of U.S. Precision Lens.

Friday, we signed an agreement that calls for USPL to be acquired by the Corning Glass Works of Corning, New York. When I last talked to you, we did not know for sure this was going to happen. In certain employee meetings I was asked about Corning, and I said we were extremely interested in Corning technology, and Corning was extremely interested in USPL technology—that we were investigating joint ventures and other relationships. I told you the truth, but I did not tell you the extent of the details covered in our discussions. The reason was very simple—legally I could not. Corning is a publicly traded company, and there are very strict Federal laws on what can be divulged and not divulged.

Under this new arrangement, USPL will operate with our present name and our present management. I will explain in detail why we have done this, what it means, and how we expect to operate in the future. At the end of these remarks I will try to answer any questions you may have about what this means.

First, I want to tell you how this came about. A few months ago we recognized that many more glass elements will be used in projection video lenses. Hybrids, that is glass/plastic combinations, are going to be more important in the years to come. We started talks with Corning on how we could use its technology to do innovative things with glass and have another American source of supply besides Schott Glass. In these discussions, it quickly became apparent that the Corning people were pursuing technology that was of high interest to USPL. We also learned that they were very interested in what USPL could do for them in a variety of their businesses. They were particularly interested in auto headlight applications, fiber optics and small aspheric lenses, as well as plastics in general.

Now I would like to talk about what Corning gets out of it and what USPL gets out of it. Corning recognizes that for many applications glass/plastic combinations, or all-plastic systems, can be better than glass. They long ago determined that USPL had ability they did not have but would be important for them to develop. For example, Corning makes all the General Motors sealed beam headlights, but, as you have seen on many new cars, headlights are now of a different design with wrap-around plastic shapes that encompass optical facets. We have ability in this area and already are working with auto companies on dashboard optics and other products. Corning is the world leader in fiber optic cable. They recognize that plastic optics will become more and more important in splicing or coupling the fiber. We, of course, have been working on plastic fiber optic coupling lenses for years, and have relationships with Hewlett-Packard, AMP and others. Corning has been working on molding small glass aspheres similar to what we are doing with plastic. They know that for many applications plastic makes more sense. Corning is a leader in TV picture tube glass. They can provide innovative faceplates for projection TV tubes that are part of the lens system optical design. We know how to integrate all areas of those designs into advantageous systems.

So then—what does USPL get out of this? There are two aspects of our business that have concerned us for a long time. First, as I said, we see glass becoming a more important factor in our projection television optics. In my opinion, Corning is the finest and most technologically advanced glass company in the world. We need a responsive, low cost American source of supply. No one can give us that better than Corning. But, beyond that, we need some clever innovation so we simply don't use glass like everyone else does. We want to develop glass aspheres that cost only slightly more than glass spheres. This would be a real breakthrough against the Japanese competition—a real technology jump. We have been working in R&D on two different programs to achieve this. Corning can help us with glass raw material and their own technological expertise.

The second aspect of our business that has concerned me is the percentage of our volume that is in projection television. While projection TV is a fantastic business that has a great future, I would still like to see us become more diversified. Corning's interest and position in the automotive market, fiber optics and glass aspherics brings us new long-term opportunity for diversification. We, of course, add to their diversification and bring new technology to them.

We expect the final close of the transaction to be September 4. After that, how will things be different? The fact is, we do not expect them to be outwardly any different than they are now. I will continue in my present position, as will the rest of our management team. I have told the people at Corning I have no personal future plans beyond USPL. My greatest desire is to see this company grow and be successful for all of us, and all of us means the people of USPL and Corning. Nothing would give me greater personal satisfaction than to look back ten years from now at a much larger U.S. Precision Lens that is better for everyone because of what we were able to accomplish as a result of the new opportunities that came from being a part of Corning.

As I said earlier, we will operate as a separate company, with a board of directors that includes USPL management people and Corning management people. We will retain the U.S. Precision Lens name. Our customers will understand that we are a subsidiary company of Corning. We all must understand that Corning is acquiring us because they like the business we have built. They don't want to change that. What they do want to do is add technology and combine efforts that allow us both to grow faster and in ways that are better than we would achieve separately. We want the same. Unless we fail to perform, the only time you will see Corning people here is when we ask them for their help and when they think they have ideas that will help us.

A bit about Corning—it was founded in 1851. It does $2 billion in annual sales, and has approximately 25,000 employees worldwide. Its headquarters are located in upstate New York near Elmira. They own companies all over the world, and have joint ventures with many more. Among the countries they have a presence in are France, England, India, Canada, Japan, South Korea, Belgium and Australia. Corning makes too many different products to mention, but you all know about Corning Ware. They also make Pyrex, laboratory glass, sealed beam headlights, automobile catalytic converters, TV tubes, fiber optics and do blood analysis.

The company is headed by James Houghton, who is slightly younger than me. He and his people have written a business philosophy that is very similar to ours. As you know, we feel very strongly about high standards. One of the most pleasing things to me was to find, after spending a lot of time with their people, that they and we have virtually identical beliefs on how a business should be run. I have no doubt Corning management believes the most valuable asset a company can have is good people. Good people thrive in an atmosphere where dignity and respect are present, and where there is opportunity for growth. Finally, I want you to know that a lot of time, investigation and thought went into this decision—not just by us but also by them. We are both very enthusiastic that this is a good thing for our companies. We have to perform as we have performed in the past. Corning is not going to give us more money, nor are they going to change the way we do things around here. They want us to keep doing what we've been doing. They will expect us to make it on our own. They know that we have talked a lot about change and that we recognize that our company has to change dramatically to be more competitive in the future. They face the same problems in all of their businesses, so they understand completely what we are talking about. What they will give us is an excellent low-cost source of raw glass, glass technology, fiber optics technology and market opportunity in new areas. All of this means that if *we*—and I underline "*we*"—get the job done, it means a brighter future for USPL and its people. Our goals are unchanged—quality up, product performance up, cost down, price down. That's how we all succeed. Now let me try to answer any questions you may have.

The due diligence process continued. On August 24, Bob Ecklin and I flew to Japan for customer visits similar to those that had been made in Chicago. We met with high-ranking executives at Hitachi, Sony, Mitsubishi and NEC. Although we were not selling lenses to Mitsubishi, the prospects for success looked favorable. The Hitachi, Sony and NEC people were very generous with their words about USPL technology and people. Each evening we had dinner with executives from one of the companies. Ecklin was particularly impressed that we had relationships at such high levels and at so many levels. He volunteered that we did a better job in this regard than Corning.

On August 29, the acquisition price was fixed at 1,166,810 shares, less the 5% that would be paid in cash. Corning stock was selling for $59.25 a share, giving USPL a value of $70,133,492 including the $1 million that would be received in five years. On that day the Dow Jones Industrial Average closed at 1898.34, and the Standard & Poor's 500 Stock Index closed at 252.93.

Jack Collins, Dave Hinchman, Don Lerner, Brian Welham and I flew to New York on September 3 for a celebration dinner the evening before the close. Roger Ackerman, Bob Ecklin, Bill Ughetta and a few other Corning people who worked on the acquisition were in attendance. Except for nervousness from the fact that Corning stock had been quite volatile and was down two points from the August 29 price fix, we were comfortable that both our companies had made an excellent deal. The next day, after anxiously watching Corning stock bounce around, we signed the final agreement. Roger Ackerman graciously gave each of us beautiful Steuben Glass mementos as we began the new relationship along with a very large stack of Corning stock certificates. U.S. Precision Lens was now a subsidiary corporation of the Corning Glass Works reporting to Ackerman. Our titles and positions remained as they had been.

Although no one was leaving, selling the company was not an easy thing to do for any of us who had been so deeply immersed in its creation. The elimination of technological risks, customer concentration risks and foreign currency exchange rate risks, from a shareholder standpoint, were the compelling reasons we did it, but there were others. Glass lens blanks were going to become a huge expense item, and Corning would provide us competitive advantage in that area where we could well have had competitive disadvantage. We thought fiber optic connectors might become a major business, and Corning's expertise in the field could help make it a reality. Their intellectual property expertise was widely acknowledged, and they had been highly successful in defending patents in Japan. Their reputation and capability in this area would surely be important to us. And finally, the sale allowed me and others to diversify our holdings and sensibly begin proper estate planning. In spite of all these good reasons, it was still a difficult decision.

A new board of directors was elected immediately after the closing. It included Jack Collins, Dave Hinchman, Don Lerner, Brian Welham, Roger Ackerman, Bob Ecklin, Jim Riesbeck, Bill Ughetta and me as chairman.

The next evening, we assembled the USPL board that had been so helpful over the years for a farewell dinner at Cincinnati's Maisonette Restaurant. Buzz Bullock, Bill Mericle and Jack Roy had significantly contributed to making USPL what it was, and we wanted to express our gratitude. After much pleasant reminiscing, we presented each with a large Steuben Glass bowl as a final thanks for what they had done.

Dick Farmer, my close friend of over thirty years, had been helpful with advice on a couple of occasions during the negotiations, and was familiar with the details. Saturday, the next morning after the board dinner, I called his home to say I was coming over for a visit. As I walked in the door I threw the large stack of Corning stock certificates to the ceiling. As they fluttered down throughout the room I said, "Before you is my life's work!" It made for a good laugh for a couple of guys who were friends from an earlier time when each of us had very little in the way of assets.

On Monday morning, as I turned into USPL's driveway, for the first time everything before me was not personally bet that day. It was a strange feeling.

# 60 1986 Continued

USPL associates accepted the sale of the company warily. It wasn't because there was anything negative about Corning, but rather an uncertainty about how the company might become different. For most, the feeling subsided quickly as they saw no change in the way the business was being operated. For our management, the major difference was the significant time being consumed in the exchange of technical information. Roger Ackerman assigned Dave Spille the job of promoting and coordinating its flow between experts in Corning and Cincinnati.

After meeting in depth with Corning's technical leader about fiber optic coupling opportunities, it became apparent that success would be far more difficult to achieve with than we had anticipated. This was because fiber technology was becoming more complex as it improved, causing ever-increasing tolerance requirements. There was good news in two areas. First Corning would be able to supply USPL glass blanks at exacting standards less expensively than we had imagined. This would result in enormous savings over what we had anticipated paying other suppliers. Second, they had a surplus Moore aspheric generating machine we could purchase at a good price. We had wanted one for years. This extremely high precision machine was capable of cutting complex curves into mold inserts at tolerance levels far more accurately than our equipment would permit. Using historic techniques, our craftsman could eventually get to the required accuracy, but it took a huge number of hours. The lead time on a machine from the manufacturer was close to a year, and the cost was well in excess of $500,000. We were able to obtain the surplus machine quickly at a significant discount. It represented an important increase in capability and tooling capacity.

It became apparent that scientists in the highly-regarded Corning Research Laboratory in Sullivan Park, could provide expertise on internal part stress and mold flow analysis as well as on many other technical questions.

Al Michaelsen, Corning's chief patent lawyer, was briefed by Dave Hinchman, Brian Welham and Opcon principals on USPL patents. We wanted his input and assistance on how to defend in the best way the company's intellectual property. He quickly became an indispensable resource, and I was soon convinced that his superb reputation, gained during the highly successful defense of Corning fiber optic patents in Japan, gave would-be violators pause as to how to deal with us. In the years to come, that conclusion would be reinforced again and again.

Finally, technical discussions were initiated with North American Philips on projection CRT faceplate curvatures to improve optic system designs.

The period from July 1 through December 31, 1986 would normally have been the first half of our fiscal 1987. Corning kept its books on a calendar year basis and to avoid reporting complexities, USPL switched to be consistent. The six-month period ending December 31 brought results similar to the same period a year earlier.

PTV sales to dealers were 304,000 sets, for an annual increase of 15%. It was a slowing from the previous year's increase of 36%, with the principal cause being disappointing December Christmas sales. By July, cumulative sales of good quality projection TVs since 1980 had exceeded one million sets. A number of makers were addressing the problems of insufficient advertising and inadequate attention to in-store displays. These were good signs for the future.

The yen to dollar exchange rate ranged from 203 to 152. At 152, the yen had finally increased in value to a level that we believed was reasonable. It meant that Japanese customers were paying 25% less for USPL lenses than they had been at its yearly low. It also meant Japanese competitors would have to sell their lenses for 25% less to maintain their relative competitive position with USPL. In the previous year the yen had fallen to a low of 263. At 152, in less than two years the U.S. dollar had depreciated 42%, requiring Japanese competitors to reduce their prices by this amount if they wanted to compete at the same U.S. dollar levels. We were amazed at how easily they seemed to make the adjustment. It caused me to wonder if, on a comparative basis with U.S. companies, they had been making far higher actual profits than were being reported in the years leading to this change. That might have been possible if capital expenditures could have been written off with greatly accelerated depreciation rates. However, we had no facts to support the premise. It was baffling. In any event, if USPL had been required to reduce its prices 42% to remain competitive, the situation would have been disastrous. We recognized currency exchange volatility as a major risk, but at last it had worked in our favor. Inflation in 1986 was 1.9%, the lowest since 1965. The prime bank lending rate dropped by 1.6% to 8.3%. Americans saved 6% of their income, which was down .9% from the previous year. Unemployment was essentially unchanged at 7%. With the exception of huge Federal budget deficits, American business people viewed the economic landscape as being very positive.

# 61 1987

In January, much to our chagrin the International Molders Union had obtained enough of our associates' signatures to force a referendum on unionizing USPL. In the previous months they had worked very secretly and selectively in pursuit of their goal. Many of the associates didn't know anything about the effort until it was announced the union was close to having the required number. We didn't know why this was happening because there were no indications of unhappiness. Some speculated the uncertainty of what Corning's ownership might bring was a factor. Prior to the 1980 vote, forced by the same union, there were inconsistencies in our administration of policies between factories and departments, so that organizing attempt was easier to understand. This one was not. The 1980 vote resulted in 87% of the associates rejecting the union. That was an overwhelming majority—far greater than is normally seen. Given that previous experience, it was surprising they would try again.

We had worked hard to be a progressive employer with clear and fair policies for associates. The culture espoused in the Business Philosophy of U.S. Precision Lens was taken seriously. To have a second referendum forced by the same union that had been so severely rejected earlier was embarrassing. Even more important to our associates and the company was the potential damage unionization would cause with customers, because in most cases USPL was a sole source of supply. If the company became unionized there was no doubt a number of customers would protect themselves by qualifying additional suppliers even if it meant paying higher prices or accepting slightly lower performance.

As before, consultants were employed to help us understand the issues. I met with all associates to hear their concerns, answer questions and give management's point of view. There were no great issues this time. The vote came about because there had been a long, secretive signature campaign conducted by a few disgruntled associates with ties to the union. They had targeted associates who joined the company after the 1980 attempt. With a representative of the National Labor Relations Board officiating, the vote took place January 16. Again, the Molders Union was rejected overwhelmingly, with 82% of associates voting no. Experts in labor relations could not understand why they tried again or why they didn't withdraw the petition when the results became predictable. That was the last time we ever heard from the Molders Union.

USPL managers and human relations executives redoubled their efforts to be sure the company not only had good policies but that they were administered consistently with improved communication. It was an area that required continual vigilance.

The LOH glass line machines began arriving in 1985. Peter Scherb had assembled his engineering and production teams. They had acquired the programmable robots that would be utilized, planned automation procedures and obtained USPL-designed conveyors that would feed raw material to the lines. The new lenscleaning system was installed, as well as a new glass thin film vacuum deposition coater. Tooling had to be built for cleaning and coating. An optical toolmaking facility was established and the associates who would operate it were trained. Our bet that USPL could make glass lenses in Cincinnati cost-effectively with shorter lead times than required by Far East vendors was about to be tested. And—there was no good alternative if we could not. It simply had to work!

To learn a complex new technology is difficult. To simultaneously automate one is extremely difficult. A production line consisted of a lens blank feed conveyor, nine different processing machines, an internal product flow conveyor and five programmable robots. Sensors were employed at virtually every point to determine if there was a process error or failure. The tooling had to be made with tolerances expressed in millionths of an inch. When a glass line operated, each machine in it had to function precisely in relation to all others. To express the requirement in analogous terms, it had to run like a Swiss watch. We had never built a Swiss watch.

The 100mm line was the first to go into production, running a very high-volume element used in a Magnavox lens assembly. We thought we had provided for an ample preproduction startup learning period. When that came to an end and the line had to produce for scheduled deliveries, it was operating anything but smoothly. There were little problems at virtually every step of the process, including tooling accuracy. Peter, his engineers and production people worked long hours identifying the exact causes and then making corrections. There was good product yield from the beginning, but at far less of a rate than the 40,000 a month goal. The problems went on for weeks as one flaw would be fixed only to have another appear. It was extremely frustrating for everyone involved, as shipping schedules were barely met for Magnavox who, at the time, was our largest customer. Making that line produce to design levels was our number one priority. If we could not, the entire effort would result in a colossal failure. To emphasize the importance of the project, I decided to temporarily move my office to the glass line control room located a few feet from each of the lines. The fact was there was almost nothing I could do to help, but everyone quickly learned how keenly concerned I was about getting the problems behind us and just how critical it was to our future. If nothing else, I thought they might cure the bugs in the line more quickly just to get me out of their presence. My stay there lasted over a month. Production gradually improved to a level that approached the design objectives, and schedules were comfortably met. My role was meaningless, except as a cheerleader. Startup of the 175mm line had similar difficulties, but the learning curve was faster.

Within a year, production goals on both lines were being met and two new lines would be ordered. Eventually we would have six of these highly automated production systems. Peter Scherb led his engineers to continue making improvements and modifications that, in time, would allow us to produce at levels far exceeding the original objectives. We had made cost-effective glass lensmaking in the United States become a reality. By combining a new technology with robotics and clever material handling, our people created the most automated large glass lensmaking process in the world. A glass lens blank at USPL, after being loaded into a feed conveyor, is never touched by a human again until it has gone through nine process steps, been placed in coating racks, been through a series of cleaning baths and received multilayers of vacuum deposition antireflective coatings. The development and successful implementation of this technology was a great achievement. It gave us competitive advantage.

In 1987 the only major PTV maker in the world we failed to do business with was Mitsubishi Electric, whose engineers in Japan preferred to buy their lenses from Japanese competitor Norita. This was significant because Mitsubishi's products enjoyed a fine reputation and were sold in high volume. Since our first trip to Japan, we had called on them relentlessly. We always were pleasantly received but never seriously considered as a supplier. This was in spite of the fact that cabinets with our designs could be made smaller and our pricing, which we were confident, was lower. USPL sales efforts with them had been very discouraging.

About that time Mitsubishi began running full-page newspaper advertisements implying they had no bias against doing business with non-Japanese companies, and encouraged outside vendors to solicit their business. After another frustratingly nonproductive visit with the Kyoto PTV engineers, I made an appointment to see Mr. Y. "Super" Yamaguchi, a very high level Mitsubishi manager in Tokyo Headquarters who once ran their U.S. operations. We had met on a couple of occasions, but I didn't know him well. He had been given the nickname "Super" during his stay in America. At a luncheon meeting that included a couple of executives reporting to him, I reviewed our history of trying to do business with Mitsubishi over the years and how successfully we had worked with all their competitors. They listened intently and asked a number of questions. Finally, I unfolded their full-page newspaper advertisement that had been running in *The Wall Street Journal* and gave it to them. After they looked at it—perhaps for the first time—I asked if this really represented their thinking or was simply a public relations effort to blunt widespread criticism of Japanese companies' lack of procurement from foreign suppliers. It was a bold and frank question given the extremely courteous way business was conducted in Japan. Yamaguchi, however, was an uncharacteristically outspoken Japanese executive, probably as a result of working in the United States, and did not appear offended. He assured me the advertisement was an honest representation of their thinking and recommended we continue our efforts to sell the people in the Kyoto Works.

It was not long after this meeting that Mitsubishi seriously became interested in USPL lenses, and soon thereafter, our products were being incorporated into a number of their models. When that happened, we could say USPL provided optics to every projection television maker in the world. Over the years Mitsubishi became one of our very large customers.

In July, Thomson, France's leading electronics manufacturer and a nationalized company, bought the RCA television business from General Electric. G.E. had owned RCA for only two years. It was hard to believe—our friends at RCA now worked for the French government. It did not seem like a good thing for the American TV industry. The transaction left Zenith as the only American-owned television maker.

An unforeseen benefit of the new relationship with Corning was the introduction of their total quality management process to USPL. Shortly after becoming chief executive officer in 1983, Jamie Houghton led an effort requiring all Corning associates to be taught the principles of total quality management. It involved learning four principles: meeting the requirements, producing error-free work, managing by prevention, and measuring the cost of quality. A number of companies were adopting similar programs based on the same techniques. After considerable study, Corning tailored its own version. The learning of a common language of quality terms and the ten actions required for success were basic parts of an introductory training program called "Quality Principles and Practices."

All USPL associates attended one of a series of week-long classes, during which the process was introduced. A follow-up one-week course entitled "Principles of a Collaborative Workplace," was also attended by everyone. Meeting the requirements of internal and external customers without error and on time, every time, were points drilled into everyone's heads over and over. Corrective action teams and quality improvement teams were formed to deal with a myriad of issues. The measurement and display of key result indicators (KRI's) on nearly every critical function became a prominent component of the system as continual improvement was pursued in all areas.

Though initially a skeptic, I soon came to believe the process could lead to an important upgrading of the company's performance. The objective was not unlike what we were trying to achieve with the "Staying Number One" program introduced in 1985. The Corning approach, however, was vastly superior because it employed proven techniques, extensive training and a common total quality language to make it a success. In retrospect, our earlier effort could have been far more beneficial if we had started with a better unnderstanding of those fundamentals.

In October John D. Rudolph joined USPL as vice-president, optical components. He had earned a chemical engineering degree from the University of Delaware and an MBA from the Massachusetts Institute of Technology before joining Corning fourteen years earlier. Well grounded in technology, and with some understanding of optics,

John was to help us look for new standard optics opportunities as his initial job. Working with Stretch Hoff, automotive, ophthalmic and universal product code reading devices were to be his first areas of concentration. Eventually, sales of code reading optics grew to be the largest business for that division. In the years to come John would become deeply involved with PTV lens customers in a variety of technical areas and would pursue the acquisition of new optical technologies to complement and extend the company's existing capability. John Rudolph has been and remains an important contributor to USPL's success.

1987 was not a good year for projection TV sales. In fact it was the only year since volume tracking of the category began that sales were down. The industry had forecast 25% growth. When the final numbers were tabulated, unit sales were actually down 4%. Inventories bulged, causing disruptive frequent changes in manufacturing schedules. Substantial amounts of 1986 model inventory were cleared at extremely low prices, dampening the sales of higher-priced current year product. Finally, 35-inch CRT sets were being heavily promoted by a number of manufacturers. Coupled with aggressive pricing, this led to direct-view sales that would otherwise have been PTV.

The PTV sales problem, combined with the costly startup of the glass lines, caused us to have a disappointing year. On a calendar year basis, sales were approximately $33 million—down $2 million from 1986. Net after-tax profit was $2.6 million—7.8% of sales. The previous year it had been $3.9 million—11.2% of sales. The standard optics division represented a bright spot with sales up 25% to $6 million.

We conservatively forecasted PTV sales for the coming year to dealers of 315,000—an increase of 7.5%. We had no doubt about PTV having a good future, but the impact of newly introduced 31-inch to 35-inch direct-view sets caused additional uncertainty in forecasting.

It had been a good year for the U.S. economy. Unemployment was down to 6.2%. The prime bank lending rate was relatively stable, ranging from 7.5% to 9.1%. Inflation was up somewhat to 3.7% but still quite low in terms of recent history. The yen/dollar relationship fluctuated from 159 to 121. We continued to be bewildered by how well Japanese exporters adjusted to the difficulties this forced upon them.

# 62 1988: A Year of Major Organizational Change

It was my eighteenth year as chief executive officer. During the previous few months I had begun thinking about management succession and personal transition that would give me more free time. My management style was to be deeply involved in the initial stages of product strategy, new manufacturing technology, capacity planning and organizational planning. There was much emphasis placed on understanding these elements so thoroughly that little was left to error once decisions were made. Thereafter I delegated freely while remaining well informed on how plans were being executed. All of our activities were driven by the desires of customers or what we anticipated would become their desires. With only ten accounting for over 80% of sales, our attention to them was intense at all levels.

With virtually every major customer, I had been heavily involved in the initial selling effort as well as the ongoing relationship along with Brian Welham, who led the way on technical matters. In the late 1970s, Dave Hinchman took charge of the standard optics business as I concentrated on Projection TV. However, PTV was growing explosively and after my first few trips to Japan he turned that activity over to Stretch Hoff, freeing himself to join us in that segment of customer relationship building and servicing. In the years ahead, manufacturing leaders Len Kosharek and Peter Scherb would also have considerable personal contact with those who bought our lenses. Dave Szkutak became part of the expanding sales team, as did John Rudolph. All external efforts were supported by a superb growing internal sales service team led by long-time associate Clarice Hoffer, assisted by Bev O'Connor, Barb Snedegar and others. By any standard, our company was extremely customer-focused. The goal was not just for customers to be satisfied with USPL performance— we wanted them to be delighted and to consider USPL as a partner in their development activity. It was an aspect of our culture that I believe was central to the company's success.

Perhaps it was a personal failing, but for all the years I had been CEO, USPL was never far from my mind. There was always more that needed to be done. Some executives are able to successfully compartmentalize their thinking and at times put business aside. It was difficult for me to do that for more than very brief intervals. Although my style was to be intensely engaged in about mid-1987, I began to notice that I was losing a little interest in the operational details of the company. Whether this was a natural progression as a result of being in the job for seventeen years, or because I was no longer the principal owner, was unclear. In any event, I did not believe it was a healthy thing for the business and it caused me to think we should consider a change at the top.

In the fall of 1987, Dave Hinchman and I discussed our future. We were the same age, which was a factor in succession planning. I asked him about his aspirations. His answer was quick and clear. He wanted to run a company as a last step in his career. It had been obvious to me for sometime that he should be my successor and I told him so, at an earlier time. If that was to be, and if he was to have enough time in the position to make his own mark on the enterprise, then I needed to step aside somewhat earlier than normal.

At the time of the sale I had informally agreed to remain with USPL for at least five years. We were now nearing completion of the second year. While Roger Ackerman was visiting my home in Cincinnati, I broached the subject. He understood my concerns and also thought very highly of Dave. He went on to say that it was entirely my call if or when a change would be made. It was understood, however, that I was under obligation to remain in a capacity that would be helpful if we made the move.

It generally is not a good idea for the old CEO to stay involved when companies change leadership. I understood the pitfalls of that very well, having seen many examples where the former incumbent was not only not helpful but undermined his successor. Dave and I had worked extremely well together as part of a team for so many years that in our case I felt that with self-imposed ground rules, it would not be a problem.

Roger Ackerman was informed on how I wished to proceed, and he endorsed the plan. The other board members were consulted and I then told Dave that on July 1 he would add chief executive officer to his title of president. I would remain chairman of the board, presiding at four quarterly meetings and continuing to be involved with major customers at high levels as well as with matters of strategy. I further told Dave my participation would be governed by his wishes, to a great extent. I decided that after the change I would never again attend an operations meeting and would be very careful not to give advice on operational issues unless it was sought. Since I wanted more "guilt-free" time for personal matters, I voluntarily cut my salary and bonus opportunities substantially. Every aspect of the transition was done at my discretion.

With so much at stake with the large PTV customers, it was decided that all would be informed of the forthcoming change through personal visits by me. We did not anticipate any particular concerns on their part, but we were not going to take any chances. The visits would give me a chance to strongly express my high level of confidence in Dave and to also point out that I was not leaving the company. While this was important to do with all the large customers, it was particularly important with those in Japan where excellent high-level relationships are critical to long term success. On April 23 I flew to Japan to meet with the top TV executives in all of our giant electronic company customers. Each expressed appreciation for personally being informed in detail before the announcement, and all were comfortable with the change.

I sent the following letter on May 23 to all USPL associates announcing the transition.

The purpose of this letter is to inform you about a major organizational change that will take place July 1 involving Dave Hinchman and myself.

As chairman and chief executive officer, I have the primary responsibility for the management of the company. Dave, as president and chief operating officer, is responsible for the day to day operations. The fact is Dave and I have worked so closely that hard distinctions between who decides what have not been made. Nevertheless, when an important decision has to be made and there is not clear consensus among our top managers, it has been my job to make the decision and direct that it be implemented.

On July 1, Dave will become chief executive officer (CEO) while retaining the title of president. I will stay with USPL and keep my title of chairman of the board. The significance of this is that as chief executive officer Dave will have ultimate responsibility for the operation of USPL.

This change is 100% my idea. It was not suggested by anyone inside our company nor at Corning. A number of years ago I told Dave that at some point I thought he should become CEO. When Corning acquired us, I informed them this was the plan. I was told it was our decision to make as Corning did not wish to interfere. Last fall Dave and I started discussing it in detail. At the end of the year I discussed it with other top managers and the date was set for July 1.

Dave has been with USPL for nearly fifteen years. During this time we and other key managers have worked as a team in building USPL. We have never had a serious disagreement in that time. I have great respect for Dave's high standards as a person, the qualities he possesses to lead the company, and his deep desire to make it a much bigger and better enterprise. He will do it.

Will things be different? Yes, of course, there will be change because all progress comes from change. Since I came to USPL eighteen years ago when it had twelve people, that has been the story—constant change. In just the last two years the change has been dramatic. Think of it— we had no LMO and no high volume glass production capability. Today these two major changes in technology have revolutionized the way we produce optics. Positive change will always be required if we are to succeed. The leadership for that will come from throughout the company, but the primary responsibility for it will be Dave's.

I will remain with USPL in a slightly less-active role. My primary concerns will be strategy, new product development and customer relations. I will do everything I can to support Dave, just as I have been supported over the years. I am not leaving the company. Shortly I will be talking to you about this and other matters. You will have the opportunity to ask me any questions you wish.

Please join me in supporting Dave Hinchman in his new role as Chief Executive Officer.

Shortly after the letter was distributed, a series of meetings were held with all associates to discuss the change and state of the business.

In the eleven years to come under Dave Hinchman's outstanding leadership as CEO, USPL would grow to become a much larger business and an important profit contributor to Corning. Among his many important accomplishments, one that particularly impressed me was a concerted successful effort to push decision responsibility to lower levels. Dave recognized that if the organization was going to continue growing this would be vital. He relinquished the CEO position in July 1999, when it was passed on to Dave Szkutak, and retired as an active member of management in January 2000.

For the year 1988, projection television sales to dealers recovered somewhat to 313,000 units—an increase of 7% over 1987. Demand, was greater, but after the previous disappointing Christmas season sales, manufacturers had been cautious in their component purchases. This led to shortages of screens, CRTs and cabinets. If the goods had been available for late season sales, we believed the product category could have been up 12% to 15%. At year-end inventories were extremely low, which indicated 1989 would be a very good year. Another favorable factor was that it was clear consumers were expressing a preference for larger screen sizes.

USPL sales rebounded 22% to $40 million, with strong contributions from all areas. Net after-tax profits were a little over $4 million, up 56%.

It was the last year of President Ronald Reagan's administration. Inflation climbed slightly to 4.1%, and the bank prime lending interest rate averaged 9.3%. The yen/dollar relationship continued to work to our favor fluctuating between 121 and 137.

The Reagan presidency began with frighteningly chaotic economic conditions and a demoralized American public. The change that took place during his eight years of leadership was phenomenal. When he entered office, the bank prime lending interest rate was an astonishing 20.4%. It dropped to a low of 8.5% in the year he departed. Inflation fell from an annual 13.5% to 4.1%. Anyone competing with Japanese companies was relieved to see the grossly undervalued yen, at 261 to the dollar, increase in value to 121 during his administration—a favorable change that could not be overemphasized for those U.S. companies affected.

And—President Ronald Reagan's White House staff would choose U.S. Precision Lens as the last company he would visit during his presidency.

# 63

### The President of the United States
### Visits U.S. Precision Lens

On Tuesday, July 26, 1988, I received a call from Ron Roberts, the executive director of the Cincinnati Business Committee, a senior group of executives of which I was a member. He said that President Ronald Reagan was coming to Cincinnati for the National Governors Conference and wished to visit a successful company that had progressive policies on drugs in the workplace. We were one of a number of firms in the area that would be considered. Shortly thereafter I was called by White House Deputy Advance Director, W. Grey Terry, who said he would like to come to our offices for a visit on Thursday at 10 A.M.

Grey Terry had been given the job of selecting a Cincinnati company the president would visit on August 8. After a half-hour session in my office during which I told him about the growth of USPL, the technology, international business and employee base, we took a tour of the three factories. Knowing he had two other companies to visit, I tried to move expeditiously. However, he did not seem pressed for time and exhibited much curiosity about the operations. When being told about our pre-employment drug screening program and associate assistance efforts, he seemingly was not as interested in the subject as Ron had led me to believe. After Grey made a phone call to the White House we went to lunch, which turned out to be quite leisurely as he continued to ask questions about our employees and international customers.

Upon returning to the office he called the White House again and resumed the questioning. In the mid-afternoon, he called the White House once more. After that I finally asked if he didn't have a couple of other companies to see. His answer was no—the president would visit USPL. He said that's what all the calls to Washington were about. I asked, "Why us?" His two-phrase answer was succinct and to the point, "Quality jobs—trade victory." During the Reagan presidency hundreds of thousands of jobs had been created but there were those that criticized this administration, saying that too many were low-skilled and low-pay. Sensitive to the issue, Grey explained USPL was a great example of a company that employed entry level people who learned a technology on the job and ultimately earned excellent pay. At the time heavy imports and a substantial unfavorable balance of trade caused some critics to push for trade barriers and sanctions against foreign competitors. Because of our substantial sales to Asian and European companies, USPL to him was a great "trade victory" story that would gain favorable press coverage. As Grey Terry departed, he said a team from the White House would arrive Monday morning, August 1, to do all the things necessary to prepare for the president's visit one week later.

Ten days earlier, Dave Hinchman, Brian Welham and Dave Szkutak had gone to Japan. Upon their return they called the office from Portland, Oregon between flights. When I gave them the news of what was about to take place they at first thought it was a joke. After relaying to them the details of the meeting with the White House visitor the day before, they realized something special was about to happen and that the next week would promise to be very interesting.

As anticipated, on Monday five members from the White House advance group, three Secret Service agents and two Marine helicopter pilots arrived to begin what would be an incredible amount of preparation for the president's visit. The advance team was led by Mary Lou Skidmore. Everything she and her people decided had to be acceptable to the Secret Service, which was of course responsible for the president's security. The pilots were there because the tentative plan was for the president to fly to USPL from the Cincinnati airport by helicopter. Dave and I, along with the USPL leadership team, met with the group for introductions and to get acquainted in the small building that served as our training center across the street from the offices. This was made available for them to use as their headquarters.

They informed us the president not only would tour the factory but also would like to address our people. We explained that actually there were three factories, and only the new McMann East facility was laid out in a manner that could accommodate a large group of visitors comfortably. It housed the four glass lens processing lines that we had at the time. These unique, highly automated production units were impressive to see in operation. Since the engineering and acquisition of the glass lines resulted in jobs being moved from Taiwan to Cincinnati, the only reversal of the trend of the U.S. optical industry going to low labor cost foreign countries, we thought the facility had a good story as well as being a logical place for the tour. It had the added advantage of a large adjacent warehouse area that could be converted into open space for seating and would accommodate the installation of an elevated platform for speakers.

The White House advance staff and Secret Service agents wanted a tour of all our facilities to determine options and obtain ideas for their planning. That took the rest of the day. The following morning after they met, Ms. Skidmore, the staff leader, informed us the president would tour the automated glass operations at McMann East and speak in the warehouse area as we had suggested. She then asked if I would take her on another tour of McMann West—the factory that housed our headquarters, tooling and molding operations.

She was particularly interested in the toolroom that contained all the large metal cutting machines used to make injection molds. As we looked at the many precision devices employed in that activity, she explained that the president would be filmed by the news media as he spoke on a stage we would need to build in the warehouse. To give it greater industrial ambiance, she wanted it displayed with equipment used in the manufacturing process. Since plastic injection molding machines were too large, she decided we should move a number of different metal processing machines from the toolroom to the McMann East stage. I could hardly believe what I was hearing. These precision machines, that she thought would look good as a backdrop, each weighed tons, and would not only require rigging specialists to move but also a massive superstructure below for support. More importantly, it would totally disrupt the moldmaking operations for at least two weeks. I resisted as Ms. Skidmore insisted. On the second day of the advance staff's visit we had already reached an impasse. I asked her to let me think about what we could do to solve the problem.

Meanwhile, with the glass lens manufacturing operation now chosen for the president's tour and the adjoining warehouse being where he would speak, Pete Cutler and his maintenance department team went to work to clear the warehouse, give it a fresh coat of paint, install a large fan in the wall to move air-conditioned air into the area from the processing room, and build the stage. Added to the tasks was the building of a platform at the opposite end of the room from the stage for press reporters and camera personnel. The platforms were quickly supplied by a company that served the local convention market. Heavy bullet-proof metal armor plate masked with blue drapery, furnished by the White House, was installed from a centered presidential podium to the edges of the platform.

We did not have flagpoles at USPL, but decided we had better get some quickly. Within three days, two large poles were installed. An American flag and USPL logo flag were quickly acquired. The second pole would be used in future years to fly the country flags of our many foreign visitors.

By Wednesday it became definite the president's helicopter, Marine One, would transport him from the Cincinnati airport to USPL. In addition, four other huge helicopters carrying the White House staff, Secret Service personnel, official visitors such as Congressmen Gradison and McEwen, and members of the press corps would also arrive in the same location. The area chosen was the large parking lot next to the McMann East factory. There was, however, one problem. The tall light poles in the lot were not compatible with helicopter blades, so they would have to come down. Within a day Pete's maintenance crew had them removed and out of sight.

The impasse on what would form the stage backdrop had not been resolved. I took Ms. Skidmore aside and informed her it would be impossible to move the toolroom machines. But, in thinking about it, perhaps there was a better idea. I told her one of the primary reasons USPL had been chosen was that we made unique technological products that were sold throughout the world. That was the story that should be told in the display. A clear simple way to express it would be to line the back of the stage with large shipping containers each with big labels indicating the various countries to which we shipped. To our great relief she agreed to the idea. Within hours the containers were assembled and labeled, "Ship to Japan, Korea, China, France, Germany, Taiwan" and other locations. It was quick and inexpensive.

It was decided that Dave Hinchman and I would greet the president at the factory door near the location where his helicopter would land. He then would meet our top management team in an adjacent room where pictures would be taken with each shaking the president's hand. Dave would show him a display of USPL lenses and products in which they were utilized and give him a tour of the glass lines during which a couple of technicians would help explain the process. Next, the president would go into a small holding room with me where he would relax as the two of us waited for the White House staff people, official visitors and press corps to move into the warehouse. When that had taken place the president and I would enter the warehouse. Once we were on the stage I was to give a short talk about USPL and introduce him. After that he would speak and depart.

As the week progressed, the visit kept getting bigger in scope. The White House staff decided a military band was needed. Soon it was arranged for a band from Wright Patterson Air Force Base in Dayton, to be on hand to provide patriotic music. It would be stationed on an open mezzanine behind the stage. More people from the

White House and the press would be coming than originally anticipated. It was further suggested that we give the president a gift. When asked for ideas the staff proposed things that had nothing to do with our business or Cincinnati. As for my introduction of the president, they suggested it be ten to twelve minutes—a length that I thought was very excessive.

We decided that for a gift we would make a gigantic magnifying glass with an elegant walnut handle. It would be mounted in a walnut frame on red velvet with a small plaque stating it was a gift from the people of U.S. Precision Lens to President Ronald Reagan August 8, 1988. The White House staff was delighted.

Meanwhile I wrote a draft of the introduction that was about ten minutes. They decided it was too long. I reminded them that from the beginning we thought ten minutes was too long. Then Ms. Skidmore said the White House would like to have two or three employees on stage with the president, and perhaps I could say a few words about them. It was a good idea. I then went about rewriting the introduction.

We wanted all associates to be able to see and hear the president, but there was a space issue. Only about 500 people could be seated in the warehouse. With over 600 associates, plus space for in excess of a hundred people from the press, official visitors, the White House staff and invited USPL guests such as suppliers and friends, we had a problem. The answer was found by giving associates a choice of being inside where they would see the president give his talk, or outside where they would see the helicopters land, the president disembark, enter the building and finally leave. A sound system would be placed outside to carry my introduction, the president's talk and music played by the Air Force band. Fortunately, about half of the associates chose the outside location. Different color tickets were printed that indicated their choice. The tickets were required at entry check-points inside as well as outside where everyone would go through metal detectors.

Choosing three associates to profile in the introduction and to be on stage with the president was a difficult selection process. There were many who had interesting career backgrounds at USPL. Finally we decided there would be one person from each factory, and each would represent a different work background. Tami Bering, a ten-year associate, joined the company in an entry level position and eventually had become a supervisor. She was chosen to represent the Bach-Buxton facility. Sixteen-year veteran Dick Wessling, our principal optician with key responsibilities at the heart of USPL's technology, was selected to represent McMann West. Jill McClure, a fourteen-year associate who had been very involved in the initial startup of the massive-optics process and had been retrained to make precision optical tooling for the glass lines, represented McMann East. The three popular choices were very excited by their roles in what was about to happen.

As we observed the White House staff and the Secret Service interact, there was obvious tension. The staff wanted the president exposed to press cameras at every turn. His visibility was their priority. With quite an opposite objective, the Secret Service wanted him publicly exposed as little as possible. It was interesting as they planned exactly where he would go and what would happen at various points. Compromises were reluctantly made between the two groups with the Secret Service generally prevailing.

As the week came to an end, USPL had gone through an incredible transformation. There was fresh paint everywhere and the factory was spotless. We had always had good housekeeping but now it was brought to a new level. And, it all happened with lightning speed. It was clear that every associate was extremely proud the president was coming to USPL, and everyone wanted the company to look its best.

Friday I received a draft of the president's speech. This was important, because I wanted my introduction to be consistent with what would follow. Saturday morning I gave the staff a copy of what I would say. It had been pared to five or six minutes and incorporated comments about the associates who would be on the stage. The person in charge of speeches said he would look at it Sunday and inform me if changes needed to be made. I told him to "read it now—because I don't do introductions of presidents with one day of preparation." He agreed, looked quickly at it and said it was excellent.

Saturday the Marines scheduled a practice fly-in and parking lot landing with two of the huge helicopters. Many of us went out to watch it. As they made their thundering descent every loose bit of sand and grit on the surface was blown into the air blasting us unmercifully. The impressive show left the lot immaculately clean as our astonished neighbors stood at their fences looking on with awe.

Monday, August 8, was a beautiful sunny day. By 8 A.M. there was a flurry of activity at USPL. Security people were on hand from the township police, sheriff's department and Ohio Highway Patrol. They were directed by the Secret Service in line with the extensive planning that had gone on during the previous week. A large fire engine was positioned on the edge of the lot where the helicopters would land. Police on horseback and on small four-wheeled utility vehicles were in the wooded area behind the factory. Secret Service marksmen armed with rifles were on the roof. An armor-plated presidential limousine had been flown in from Washington in case it would be needed. There were press people everywhere. The major television networks, CNN and all local TV stations had reporters and camera operators present. We never heard an exact count, but know there were well in excess of 150 people from the press in attendance. They had assigned areas both inside and outside the building. A small pool of reporters from the White House press corps was selected to witness the president's tour of the factory, and they would share what was observed with the others. Before anyone was admitted to the factory, Secret Service agents with specially trained dogs went through the entire premises in search of explosives. A couple of our raw materials contained similar ingredients, causing the dogs to be a little nervous, but in the end the agents were satisfied the building was safe.

As people filed through the roped entry points showing their tickets and passed through metal detectors, the Air Force band played John Philip Sousa marches that could be heard at both the inside and outside locations. The atmosphere was very festive. I explained to everyone in the warehouse exactly what would happen because they were going to have some waiting time after the helicopters landed and the president was touring the factory. I then went over to Gerry Miller, whose firm built the facility, to tell him he had a slight chance of becoming the most well known contractor in America. He asked why. I said, "If any of the walls come down as a result of the five helicopters coming in, everyone in the world is going to know your name!"

At about 10:30, on schedule, we could hear the roar of the five helicopters that were about to descend into the parking lot. The last to arrive was Marine One, carrying the president. The press corps stationed near the building received an even bigger sand blasting than we did on Saturday. We later learned it was one of the hazards of being assigned to presidential trip coverage. Marines in full dress uniform came out of the back of Marine One and opened the front door which carried the Presidential Seal. They stood saluting as President Ronald Reagan, with his great smile, came down the steps waving to our associates outside. As he walked towards the factory door, the armored presidential limousine slowly moved along, shielding him from the outside crowd with one door slightly open in case there was a need for a quick getaway. The president had already survived one assassination attempt, and the Secret Service was not leaving the slightest detail to chance.

As Dave and I met President Reagan at the door, I was struck by how much older he looked than I had envisioned. After introductions, we moved into the room where he would meet and be photographed with Jerry Behne, Jack Collins, Pete Cutler, Stretch Hoff, John Rudolph, Peter Scherb and Brian Welham. From there it was on to his tour of the glass lines conducted by Dave. While that was taking place Brian and I showed the Congressmen and senior White House staff, who traveled to Cincinnati with the president, the lens and product exhibit.

The White House advance staff had one overriding objective —a "great photograph"—one so good that it would be carried in papers everywhere. The one they planned was one of the president looking through a big circular coating rack that held 160 glistening large lenses, each secured with a series in intricate clamps. It would indeed make for a dramatic picture. The coating rack was placed between two glass lines. One of Dave's jobs was to guide the president into the proper position for the photographers, and he did it precisely. The advance team was ecstatic—they had their "great photo." When Dave completed the tour the president and I went into a small first aid room that had been outfitted with a comfortable chair, bottled water and even paper coasters with the Presidential Seal. As we waited and chatted, everyone else moved into the warehouse that had been temporarily converted into an auditorium.

A few years earlier Joyce and I had built a home theater. We decorated the walls with a half a dozen old movie posters. One of them was for a movie called *Desperate Journey*, starring Errol Flynn and Ronald Reagan. The 1940s poster carried both their pictures in an action scene. With prior staff approval, I hung it on the wall in the holding room without the glass. He had been told we would like him to sign it. When he saw it he said, "Wow—that was a long time ago." He then wrote on it "To Joyce and Roger with very best wishes— Ronald Reagan." A photographer peeked in the door and snapped a picture of him as he signed it.

After about five minutes one of his aides came in to say that everyone was in place and we could proceed to the stage. As we approached the door to the warehouse the Air Force band played "Ruffles and Flourishes." This was followed by an announcer saying, "Ladies and gentlemen—the President of the United States accompanied by Mr. Roger Howe." Then, as the band played "Hail to the Chief," the president led the way as we entered to a standing thunderous applause. The outpouring of genuine good feeling towards him was a wonderful spectacle to watch. As he climbed the few steps to the stage with his famous smile and waving to the crowd, I was again struck by his appearance, but now it was how much younger he looked than when Dave and I met him at the door only twenty minutes earlier. He shook hands with Dick Wessling, then Jill McClure and, lastly, Tami Bering. As I stood at the presidential podium waiting to make the introduction, the wild applause continued on for him until he signaled it should end. I then made the following introductory comments:

> Mr. President, we all are enormously honored that you would choose to visit our Company. This is our greatest day. More than that, in the broader sense, it is a celebration of American free enterprise because it wasn't many years ago that we only had twelve employees in an area a fraction of the size of this room.
>
> That was in 1970, and since then our people have created technology that has allowed us to win business throughout the world. Today, nineteen percent of what we make is exported. Sixty-eight percent is sold to foreign-owned companies, with much of the selling taking place in their homelands, even though the lenses may be shipped to U.S. locations.
>
> Over the years, our employees have been told that we can't compete with people in the Far East or in other low labor cost areas. We were told we should license our technology because competition would quickly figure it out and copy what we do. And finally, we were told that the Japanese companies would never buy from a small company on the outskirts of Cincinnati, Ohio. Mr. President, we are very proud to tell you that we rejected all of those arguments and that is why this Company has been able to grow, create quality jobs and consistently be a fair-share taxpayer.
>
> During the last eight years, a period we thought might be of particular interest to you, we have had by far our greatest growth. This has

happened because our government has helped us the best possible way—by providing a non-intrusive free market environment in which we have had the greatest possible opportunity to prosper or fail, on our own. We thank you for all that you have done to provide that very special environment.

Mr. President, you have come here to see our people. Three of them, representing different factories, have joined us here. I wish there were time to tell you about everyone. To my right is Tami Bering, who was unskilled when she joined us in 1978. She is now a supervisor. To your left is Dick Wessling. He gave up a good job in 1972 to take a chance on what was then a tiny company. Today he is our principal optician. To your right is Jill McClure. She came at entry level in 1974 and has recently been retrained to make precise optical tooling. These people, like all of us at USPL, take part in a continuing Total Quality Training Program.

You have just seen the newest and most automated large glass lensmaking factory in the world. It represents much more than state-of-the-art optical manufacturing technology—it represents moving jobs back to America. In 1986, we sold a superb lens factory in Taiwan where labor rates were one-sixth of ours. We did it because we believed that with advanced technology and dedicated USPL people, we could eventually produce lenses at lower cost right here in Cincinnati. Our people have already proven that we made the right decision, even though it was contrary to all popular advice.

Mr. President, no one knows better than you that people only invest in the future when they believe there is a good future. Over the last eight years your policies and leadership have given us the confidence to make that investment. The new job creation you have just seen is a direct result. Again, we thank you.

To all of you outside, and all of you before me here, it is my great honor to introduce to you the President of the United State of America.

The president then rose to another enthusiastic prolonged applause. With a few ad-libbed departures from the advance script, his remarks were as follows:

Thank you very much—and thank you Roger Howe and David Hinchman—I brought a couple Washingtonians with me— Congressman Gradison and Congressman McEwen. You know if I didn't get out of Washington often, it would be easy to lose touch with what's going on. Back at the airport someone asked me about my impression of the Reds' manager, and I told him I didn't know if he was talking about Pete Rose or Gorbachev.

I can't tell you how good it feels to come here to the heartland where America's work gets done, and to get away from that puzzle palace on the Potomac. Every time I leave Washington to travel around the country, as I get out of the plane I half expect to see a sign waiting for me saying, "Welcome to America."

I came here today to tell you that you are part of a remarkable American success story. Around the country, companies like U.S. Precision Lens are leading America's economic expansion and manufacturing boom. In fact, more good news came out last week: U.S. factory orders have grown at their highest rate in eighteen years—and production levels are pushing factories to near capacity. Not surprisingly, employment is at its highest level in history. Just last month, America created another 283,000 new jobs—that makes it over 17 million jobs since our expansion began. And like yours, these are high-wage jobs with a future.

America is in the longest peacetime economic expansion on record —and it's our exports and our manufacturing that are now driving it. In many ways all of you already know this, because you've been part of it right here. Since 1980, your company has tripled sales, doubled employment, and your exports have exploded into world markets. And talk about beating the pants off the competition: you sell over a third of what you make to companies from Japan.

You're not alone. Americans today are selling shoes to Italy, medical equipment to Japan, and machine tools to West Germany. A furniture company in St. Louis now exports to Europe, its owner having discovered, in his words, "that my products were a lot better and a helluva lot cheaper." One company up the road in Columbus has even figured out a way to sell sand to Egypt. It mines and refines a high tech silica sand that is used as a cleansing agent in furnaces. From software to sand, from jumbo jets to precision lenses, American products are the finest in the world, and we can out-compete any country on Earth.

I've heard some people bad-mouthing our economy recently. I think they must have stopped reading the newspaper that day in 1981 when they handed over the lease to the White House. Yes, things were bad back then. But today unemployment is at its lowest level in fourteen years. Inflation is low and under control. America's manufacturing productivity has soared 4.3 percent a year, the highest rate since World War II. Real family income is up. Exports are at an all-time high. And America has created, on average, a quarter-of-a-million private sector jobs each month for 68 consecutive months. Any way you slice it, America has taken the pennant and is sweeping the World Series.

Now, some people are telling you to take for granted the economic growth of today and of the last seven years. Their message is, "You can take prosperity for granted, it's time for a change, take a chance on us." That's sort of like someone telling you that you've stored up all the cold beer you could want, so now you can unplug the refrigerator. But, no more than with a refrigerator, you can't unplug our pro-growth economic policies and expect things to stay the same.

Well, the fact is the whole world is learning from our example and turning away from decades-old policies of government-mandated economic failure and turning toward the type of economic policies that Vice-President Bush and I have put into practice over the last seven years. The policies that pulled America out of economic stagnation, rising unemployment, declining family income, and double-digit inflation have made America's economy a global success story. These policies are the wave of the future. Country after country is reducing taxes, cutting regulation, reducing the role of Government, and letting entrepreneurs and working people build new factories, new jobs, and new futures for themselves and their families. It's sweeping the world, but like hamburgers and baseball, it all began right here in the U.S.A.

You know—I have to tell you there is a thing called the Economic Summit every year—seven countries—we go from one country to another every year where that country's head of state is the chairman. I was the new kid in school when I went to my first one in Canada, and for a little while I stayed silent at these meetings—then our economic reforms began to take hold and what a thrill it was—the new kid walked into the meeting and the other six heads of state said, "Tell us about the American Miracle." Well, I was very pleased to do that.

You know there is a story about a fellow that was always asking Abraham Lincoln to give him a government job. One day the news was the customs chief had died. Sure enough, this fellow showed up and asked if he could take his place. President Lincoln said, "It's all right with me if the undertaker doesn't mind."

Well, no bureaucrat, politician, Government expert, or certified genius sitting in a Federal office in Washington, D.C. has ever been able to replace the economic miracle of free men and women working with their hands, their hearts, and their heads to build a better future for their families and a stronger economy for America. I have said this again and again, and I'm going to keep on saying it: It's not the Government, it is the American people who have made our Nation the greatest country on earth. Basically what our program did was get out of your way and let you do what you can do so well.

I can't think of any part of America where that's truer than here in the Midwest. You know, I get a little tired of hearing Cincinnati and other Midwestern cities called the Rust Belt when the Midwest is the heartland of America's industrial renaissance. The Midwest isn't the Rust

Belt; it's the Boom Belt. I can't help wondering if maybe Precision Lens can help out some of our critics. I think they could use a pair of high-quality lenses, because they've been looking at the world through mud-colored glasses much too long.

Well, there's another area where the example set by Precision Lens is crystal clear: I'm talking about your important efforts toward a drug-free workplace. Through pre-employment testing, employee counseling and treatment, you've really made a difference. Here, and around the country, workplace drug programs have brought dramatic improvements in worker safety. There is no place for illegal drugs in the workplace or anyplace else.

I believe that programs like yours make a positive impact throughout the community. In addition to making this plant safer and more pro-ductive, you can also be proud that you're sending a message to our children to be drug free, because illegal drug use will not be tolerated.

I'm glad you all were able to get into work today. They weren't going to let me in but luckily someone recognized me. Of course, it is always nice to be recognized. You know years ago, after a quarter of a century in the picture business and a number of years on television, you are used to being recognized. I was walking down Fifth Avenue in New York one afternoon and suddenly a fellow about thirty feet ahead of me stopped and pointed and said, "I know you—I know you—I see you in pictures and on the television screen all the time." Well, you know New Yorkers—they all stopped, everybody lined up and made an alley—here he came down the middle of the alley and he was fum-bling in his pocket. He keeps on coming telling me how well he knew me and how often he has seen me, and he sticks out a pad and pencil and says, "Ray Milland." So I signed "Ray Milland"—there was no sense in disappointing him.

It's been a real joy for me to see the work that you do and the tremen-dous pride with which you do it.

Throughout his talk, there were many interruptions for applause. He clearly was enjoying the event. At the conclusion I presented him with the frame-encased large magnifier saying, "Mr. President—we thought you might like to use this with your critics who can't seem to understand how good things are in America." As he accepted it, he turned to Congressmen Gradison and McEwen and quipped, "Now maybe I can read the fine print in those bills you send me." Then, with the Air Force band playing patriotic music, and the audience giving deafening applause, the president, Dave and I departed the warehouse. His top aides and Secret Service agents led him to the door he had entered earlier for departure. As we went outside the large crowd behind the fences who had heard everything that took place inside continued the applause. As we shook hands he thanked us, wished us well and, waving to everyone, headed for Marine One. The four other helicopters were quickly filled. Marine One was the first to lift off, with the president continuing to wave through its large window. In slightly over an hour, the visit was over and he was on his way to downtown Cincinnati to address the National Governors Conference.

Once all the helicopters had departed, everyone inside was invited to take the same tour the president had just gone on. The Air Force band continued playing for a while longer. Press reporters, and TV news reporters with cameramen, interviewed USPL associates for their reaction to the visit. The comments were universally favorable not only for the president, but also for USPL as a place to work. There was not one sour comment or regrettable incident. Our associates could not have been more exemplary in their conduct.

The next day the front page of *The Cincinnati Enquirer* carried the "great photo" of the president looking through the coating rack with Dave at his side. A similar "great photo" of the president only was seen in nearly every major newspaper in the world. The coverage was incredible. Doug Edmondson, a USPL quality technician who knew nothing of the visit because he was on an extended trip to Taiwan, was shocked to read all about it in the *China Post*. Our Japanese customers saw the photo and read the story in their papers.

The *Clermont Courier Press* printed a glowing editorial. Excerpts are as follows:

> It's not every day a president visits Clermont County. Since President Ulysses S. Grant found his way to Point Pleasant by accident of birth, only three Presidents have found their way here on purpose.
>
> More important is why the president came here. U.S. Precision Lens is a great example of what the president says his administration is accomplishing. The company is a Reagan-era Horatio Alger story—tripling sales in the last eight years, competing with and beating the Japanese.
>
> U.S. Precision Lens is, above all, a Clermont County success story. The plastic optics company had all of twelve employees when it moved from downtown Cincinnati in 1971. Now it has 600. Most of all, the visit is a credit to the Clermont County workforce at U.S. Precision Lens. Without them, none of this would ever have happened.

Many associates took video pictures of the event both inside and outside the building. Jerry Behne, and others in the Human Resources Department, borrowed them for duplication. Copies of news tapes from Cincinnati's three TV stations were also obtained. From these materials we had a professional video company prepare a tape of the entire event. Two aspects of it amused me. The first was that President Reagan, who I initially thought appeared so old when we first met, looked years younger when seen on the tape. Whether it was his training as an actor or something else, on stage and before the camera he didn't look anywhere close to his advancing years. His story about the great 1940s and 1950s character actor Ray Milland was funny for the older people in the room who recalled Milland's motion pictures, but there were many quizzical looks on the faces of our younger associates who had never heard of him. Although they laughed along with the rest of us, they missed the point that the president was mistaken for another very popular actor of the day.

The White House advance staff was delighted with everything about the event. For them, as well as USPL, it could not have gone better. In appreciation, they gave the people that were of particular help to them small gifts such as tie clasps and cufflinks bearing the Presidential Seal. Then, it was off to some new location for them, for the next presidential visit.

Our associates glowed with pride that the President of the United States would visit USPL. It also made a great impression on customers who, in virtually every case, wanted to hear the details of what took place that day.

We decided to create a commemorative booklet covering all aspects of the event for the associates and friends that were in attendance. It was called: *A Day to Remember—August 8, 1988.* It was completed nearly a year later. This beautifully-crafted piece featured a wonderful cover photo of the president smiling and waving to our associates.

On August 1, 1989, about seven months after he left office, I sent a copy to him at his Los Angeles office along with the following letter.

Dear President Reagan:

Nearly one year ago, you gave the people of U.S. Precision Lens and our community a great honor by visiting us.

To commemorate that occasion for our employees and our customers throughout the world, we recently produced the enclosed booklet. I thought you might enjoy having it and recalling the hour you spent with a 100% Ronald Reagan crowd.

I cannot let this opportunity pass without again saying how grateful we are to you and Mrs. Reagan for all that you accomplished in your presidency. What we so proudly showed you in Cincinnati was, in a large part, the direct result of all you did to revive our economy, our spirit, and our country's position in the world. You have been a wonderful partner.

With warmest personal regards.

Roger L. Howe

A couple of weeks later I received the following hand-addressed and hand-written letter from him.

RONALD REAGAN

Dear Roger

Just a line to thank you for that reminder of a "Very special Day." Believe me it was a special day for me and one I'll long remember. It will be easier remembering with that handsome brochure. Thanks too for your generous words.

Very Best Regards — Ron

*Dave Szkutak joined USPL in 1984 as a production manager. He would eventually become President and Chief Executive Officer.*

*McMann East 72,000 square foot factory completed in late 1985.*

*Dr. Akio Ohkoshi, co-inventor of Sony's Trinitron television picture tube and a helpful friend who took an early interest in USPL.*

*Unique automated glass lens processing lines developed by USPL.*

*The last gathering of the original USPL Board of Directors September 5, 1986. From left to right: Brian Welham, Dave Hinchman, Bill Mericle, Buzz Bullock, Don Lerner, Roger Howe, Jack Roy.*

*Vacuum deposition anti-reflective coating rack, controls and chamber.*

*Projection-television lens assembly section.*

*President Ronald
Reagan arrives in the
USPL parking lot
aboard Marine 1 on
August 8, 1988.*

*Dave Hinchman shows
the President a display
of USPL products.*

The "Great Photo" that was so important to the White House staff. As they had hoped, it was seen in newspapers throughout the world.

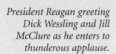

President Reagan greeting Dick Wessling and Jill McClure as he enters to thunderous applause.

*Roger Howe presenting
President Reagan with a
gift. He quipped to the two
congressmen present: "Now
maybe I can read the fine
print in those bills you
send me."*

*The Emmy.*

1997 Emmy Awards Ceremony in New York – Dick Wessling, Dave Hinchman, Ellis Betensky, Henry Kloss (who also received an Emmy), Roger Howe Don Keyes and Brian Welham.

Roger Howe accepting the Emmy on behalf of all the associates of USPL.

USPL McMann West headquarters and factory in foreground, and McMann East on far side of road in 1998.

# 64 Concluding Thoughts

Enough cannot be said about our people. In instances too numerous to relate, uncommon extra efforts on the part of dedicated individuals resulted in meaningful achievements. Behind every operation, technical development and staff activity, there are stories about associates who did things beyond what was normally expected. Because there are so many, regrettably they cannot be told in these pages without making omissions that would be inappropriate. However, collectively they made a major difference in the success of the company.

A human phenomenon not unique to our company but one that indeed existed for most of the company's long-term associates was the almost "second family" feeling that grew out of working together. It was different from the considerable pride that most associates held for USPL as a place to work. Genuine good friendships would cause joy for many because there was good fortune for one, and there would be sadness for many because there was difficulty for one. It could be seen as they celebrated achievements together and supported each other in tragedy. Virtually every associate that had been with the company for any length of time developed what they considered close friendships with other USPL people. I certainly did, and like so many others count these people as friends forever. It was a nice part of our culture to witness and a happy byproduct of being part of the enterprise.

In the preface, I wrote about the importance of a fundamentally good idea in the creation of a business. It cannot be overemphasized. Herman Buhlmann's idea for a unique watch crystal grinding machine became the foundation on which our company was founded. For the first twenty-five years of its existence, his idea motivated customers to be attracted to its products. His son Henry, a brilliant engineer, recognized that beyond plastic watch crystals it would be possible to utilize injection molding technology to make far more complex lenses. Although the idea led only to modest success for the next fifteen years, it saved the business. The problem under Henry Buhlmann's ownership was a combination of lack of technical advancement due to limited resources and perhaps, more importantly, the inability to find a way to successfully market the benefits of the technology to a wide range of potential plastic optic users. Nevertheless, it was Henry's development that I recognized as a fundamentally good idea in 1970. The promotion, development and concentrated support of that idea positioned us to take advantage of the opportunities that were soon found at Polaroid, all the semiconductor companies that made LED hand-held calculators, Honeywell with its auto-focus camera modules and the myriad of instrument applications that would later develop. By significantly advancing the technology with strategic technical as well as manufacturing capacity improvements, and finding ways to inform hundreds of thousands of engineers in many industries throughout the nation about our plastic optic capability, we were able to build swiftly upon the idea.

Henry Kloss's fundamentally good idea for the creation of modern-day projection televison was severely hampered because there was no practical technology to make the unique aspheric optics that would be required for the refracting lens system. The solution to the problem of how to make large aspheric plastic lenses practically was our one fundamentally good idea under my years of leadership. It allowed Ellis Betensky, his Opcon associates and Brian Welham to collaborate in the design of innovative optical systems that were a major departure from anything that had ever been made before. Through relentless design and manufacturing improvement, we would become lens suppliers to every major television maker in the world. It would eventually account for over 90% of the company's business.

To this day, I believe USPL has prospered because of just three fundamentally good ideas—Herman Buhlmann's watch crystal grinding machine, his son Henry's injection molded plastic optics and our invention of a process to manufacture unique aspheric optics. These have propelled the company for its first seventy years, and will propel it for many years to come.

At the time I purchased U.S. Precision Lens in 1970, there were other companies that made plastic optics. Polaroid and Kodak had substantial operations solely for the support of their own production lines. American Optical had a small division that offered custom plastic lenses to the open market. Although their volume was greater than ours, it was still quite small. Combined Optical Industries in England was larger than USPL, and eager to increase its minimal base of business in the United States. As we quickly grew to be the largest producer, that success attracted others to the business. However, no competitor ever developed the broad range of capability that we possessed. Polaroid and Kodak would later offer plastic optics to the open market without success.

I have always believed that we became the dominant maker of plastic optics because we made it our only business. By acquiring support capabilities such as internal tooling capacity, vacuum deposition coating technology and outstanding optical engineering, as well as optical design skills, we were able to make our niche in the optical industry more and more specialized. Adhering to the discipline of being only an optics company that specialized in the use of plastics was the key to what was achieved.

For those of us in leadership positions at U.S. Precision Lens, as well as for many of our long-term associates, there were many reasons for satisfaction. That we could grow a twelve-employee company that was so small and in such terrible financial condition in 1970 to a level that would bring excellent careers to so many of us, as well as provide hundreds of millions of dollars in wages for thousands of others, was high on that list. As the economic benefit of it all was magnified into our community, the effect was substantial. Another aspect of the experience that I highly value was all that we learned and the enriching friendships that developed through the opportunity of doing business with people from different cultures in so many foreign countries. This not only brought significant benefit to the company, it also gave those of us engaged in the relationships greater personal dimension. At one time or another during these years, we were actively involved with customers in nineteen nations around the world.

And—we appreciated the recognition and accolades. Great companies had given us a multitude of supplier excellence awards. In 1997 the Optical Society of America awarded its prestigious David Richardson medal to Brian Welham for having provided the technical leadership in pioneering the development and commercialization of aspheric lens technology. Also that year, USPL was a recipient of an Emmy Award from The National Academy of Television Arts and Sciences at an elaborate New York award ceremony. The technical committee that made the selection inscribed it as follows:

1996–1997 OUTSTANDING ACHIEVEMENT FOR THE SCIENCE OF ENGINEERING DEVELOPMENT HONORS
US PRECISION LENS, INC.
"FOR ITS DESIGN AND DEVELOPMENT OF HIGH EFFICIENCY OPTICS WHICH MADE POSSIBLE THE GROWTH OF THE COLOR VIDEO PROJECTOR INDUSTRY."

And finally, in 1998, I was inducted into the Cincinnati Business Hall of Fame at a ceremony where I had the opportunity to say that the honor properly belonged to all the people of U.S. Precision Lens.

U.S. Precision Lens is a wonderful company with a promising future. It will continue to bring benefit to its customers, its people, its owners and its community. Those of us that led the effort to build it were, indeed, blessed with the opportunity to—

do interesting things

with interesting people

in interesting places.

# Acknowledgments

The writing of this book required considerable research. I thank Jerry Behne, Jack Collins, Eric Grothaus, Dave Hinchman, Stretch Hoff, Don Keyes, Jim Madden, Lynn Miller, Peter Scherb, Brian Welham and Dick Wessling for furnishing me historical information and their recollections. Joseph Warkany, the grandson of the founder, kindly provided information on the Buhlmanns. Clarice Hoffer did time-consuming extensive file searches that were critical to the effort. Ed Howe compiled the historical economic statistics.

As a result of their manuscript reviews, I am grateful for the excellent suggestions of my friend, Tom Laco, and Miami University faculty members Dr. Kate Ronald, Howe Professor of Written Communication, and Dr. David Rosenthal, Professor of Marketing.

The generous Foreword by my friend of forty-five years, Dick Farmer, is most appreciated.

And, I particularly thank my long time assistant, Flo Isaac, who typed and retyped the manuscript countless times as it went through my many revisions.

# Appendices

# Appendix 1

| Year | U.S. Bank Prime Lending Rate* | | U.S. Annual Inflation Rate | U.S. Average Unemployment Rate |
|------|------|------|------|------|
| | Low | High | | |
| 1970 | 6.9% | 8.5% | 5.7% | 5.0% |
| 1971 | 5.3 | 6.3 | 4.4 | 6.0 |
| 1972 | 4.8 | 5.8 | 3.2 | 5.6 |
| 1973 | 6.0 | 9.9 | 6.2 | 4.9 |
| 1974 | 8.9 | 12.0 | 11.0 | 5.6 |
| 1975 | 7.1 | 10.1 | 9.0 | 8.5 |
| 1976 | 6.4 | 7.3 | 5.8 | 7.7 |
| 1977 | 6.3 | 7.8 | 6.5 | 7.0 |
| 1978 | 7.9 | 11.6 | 7.6 | 6.1 |
| 1979 | 11.5 | 15.6 | 11.3 | 5.9 |
| 1980 | 11.1 | 20.4 | 13.5 | 7.2 |
| 1981 | 15.8 | 20.5 | 10.3 | 7.6 |
| 1982 | 11.5 | 16.6 | 6.2 | 9.7 |
| 1983 | 10.5 | 11.2 | 3.2 | 9.6 |
| 1984 | 11.0 | 13.0 | 4.3 | 7.5 |
| 1985 | 9.5 | 10.6 | 3.6 | 7.2 |
| 1986 | 7.5 | 9.5 | 1.9 | 7.0 |
| 1987 | 7.5 | 9.1 | 3.6 | 6.2 |
| 1988 | 8.5 | 10.5 | 4.1 | 5.5 |
| 1989 | 10.5 | 11.5 | 4.8 | 5.3 |
| 1990 | 10.0 | 10.1 | 5.4 | 5.6 |
| 1991 | 7.2 | 9.5 | 4.2 | 6.9 |
| 1992 | 6.0 | 6.5 | 3.0 | 7.5 |
| 1993 | 6.0 | 6.0 | 3.0 | 6.9 |
| 1994 | 6.0 | 8.5 | 2.6 | 6.1 |
| 1995 | 8.5 | 9.0 | 2.8 | 5.6 |
| 1996 | 8.3 | 8.5 | 2.9 | 5.4 |
| 1997 | 8.3 | 8.5 | 2.3 | 4.9 |
| 1998 | 7.8 | 8.5 | 1.6 | 4.5 |
| 1999 | 7.8 | 8.5 | 2.2 | 4.2 |

* Rounded

# Appendix 2

## Japanese Yen—U.S. Dollar Exchange Rate

| Year | High | Low |
|------|------|-----|
| 1970 | 360 | 360 |
| 1971 | 358 | 315 |
| 1972 | 315 | 294 |
| 1973 | 303 | 254 |
| 1974 | 305 | 274 |
| 1975 | 307 | 285 |
| 1976 | 306 | 286 |
| 1977 | 293 | 238 |
| 1978 | 242 | 177 |
| 1979 | 251 | 195 |
| 1980 | 261 | 203 |
| 1981 | 246 | 199 |
| 1982 | 278 | 219 |
| 1983 | 247 | 227 |
| 1984 | 252 | 223 |
| 1985 | 263 | 200 |
| 1986 | 203 | 152 |
| 1987 | 159 | 121 |
| 1988 | 137 | 121 |
| 1989 | 150 | 124 |
| 1990 | 160 | 125 |
| 1991 | 142 | 125 |
| 1992 | 135 | 119 |
| 1993 | 126 | 101 |
| 1994 | 113 | 97 |
| 1995 | 105 | 81 |
| 1996 | 116 | 103 |
| 1997 | 131 | 111 |
| 1998 | 147 | 114 |
| 1999 | 124 | 102 |

# Appendix 3

| Year | Dow Jones Industrial Stock Average* | | | Standard & Poor's 500 Stock Average* | | |
|---|---|---|---|---|---|---|
| | High | Low | Year-End Close | High | Low | Year-End Close |
| 1970 | 842 | 631 | 839 | 93 | 69 | 92 |
| 1971 | 951 | 798 | 890 | 105 | 90 | 102 |
| 1972 | 1036 | 889 | 1020 | 119 | 102 | 118 |
| 1973 | 1052 | 788 | 851 | 120 | 92 | 98 |
| 1974 | 892 | 578 | 616 | 100 | 62 | 69 |
| 1975 | 882 | 632 | 859 | 96 | 69 | 90 |
| 1976 | 1015 | 859 | 1005 | 108 | 91 | 107 |
| 1977 | 1000 | 801 | 831 | 107 | 91 | 95 |
| 1978 | 908 | 742 | 805 | 107 | 87 | 96 |
| 1979 | 898 | 797 | 839 | 111 | 96 | 108 |
| 1980 | 1000 | 759 | 964 | 141 | 98 | 136 |
| 1981 | 1024 | 824 | 875 | 138 | 113 | 123 |
| 1982 | 1071 | 777 | 1047 | 143 | 102 | 141 |
| 1983 | 1287 | 1027 | 1259 | 173 | 138 | 165 |
| 1984 | 1287 | 1087 | 1212 | 170 | 148 | 167 |
| 1985 | 1553 | 1185 | 1547 | 212 | 164 | 211 |
| 1986 | 1956 | 1502 | 1896 | 254 | 203 | 242 |
| 1987 | 2722 | 1739 | 1939 | 337 | 224 | 247 |
| 1988 | 2184 | 1879 | 2169 | 284 | 243 | 278 |
| 1989 | 2791 | 2145 | 2753 | 360 | 275 | 353 |
| 1990 | 3000 | 2365 | 2634 | 369 | 295 | 330 |
| 1991 | 3169 | 2470 | 3169 | 417 | 311 | 417 |
| 1992 | 3413 | 3137 | 3301 | 441 | 395 | 436 |
| 1993 | 3794 | 3242 | 3754 | 471 | 429 | 466 |
| 1994 | 3978 | 3593 | 3834 | 482 | 439 | 459 |
| 1995 | 5216 | 3832 | 5117 | 622 | 459 | 616 |
| 1996 | 6561 | 5033 | 6448 | 757 | 598 | 741 |
| 1997 | 8259 | 6392 | 7908 | 984 | 737 | 970 |
| 1998 | 9374 | 7539 | 9181 | 1242 | 928 | 1229 |
| 1999 | 11497 | 9121 | 11497 | 1468 | 1212 | 1469 |

*Rounded daily closing

# Index

Behne, Jerome J. 190, 218, 284, 285, 309, 357, 366, 380

Bell & Howell 78, 124, 155, 202

Bendix Corporation 14

Bering, Tami 355, 359, 360

Best, John 254, 255, 256, 257, 259, 261, 262, 271, 272, 274

Betensky, Ellis 52, 53, 57, 122, 132, 149, 151, 182, 227, 260, 269, 288, 290, 376

Bingham, Pete 156

Bloom, John 137, 138

Bohache, James 221, 252

Boling, Lawrence H. 7, 9

Bosch, Robert 229, 254, 255, 259, 261, 263, 266

Boudinot, Donald 6

Bowmar 54, 59, 60, 61, 62, 80, 84, 85, 86, 89, 91

Brand Studios 16, 37

Brandinger Dr. Jay 152, 155

Buhlmann, Henry xvii, 12, 15, 17, 23, 24, 46, 245, 376

Buhlmann, Herman Louis xvii, 12, 15, 18, 194, 375, 377

Bullock, John M. (Buzz) 96, 97, 331

Bullock, John R. 57

Bulova Watch Company 12

*Business Week* 59, 85, 154

Byrd, Ronald 34, 39, 64

**C**

Campbell, Robert 23, 24, 28, 32, 41, 55, 60, 72, 83, 95, 119, 130, 133, 134

Campbell, Van 319

Canon 226, 257

Carlson, Dr. Gary 125

Charan, Ram 308, 309, 310, 311, 316

*China Post* 366

Christopher, Brian 99

**F**

Fairchild Camera and Instrument 59, 60, 61, 63, 89

Farmer, Richard T. 8, 11, 26, 331, 380

Flynn, James 320, 321

*FOCUS* 240

*Forbes Magazine* 138

Freeman, Ken 320, 321

Fujinon 288, 289, 291, 292, 307

Futurevision 136

**G**

Garrett, Wayne 45, 47, 60

Garrity, Norman 316

General Electric (G.E.) 12, 13, 121, 125, 126, 127, 128, 129, 151, 152, 153, 154, 155, 167, 170, 174, 190, 194, 201, 208, 217, 246, 248, 267, 310, 339

General Motors 134, 290, 326

Gerding, Larry 297, 301

Gifford, Dr. Jack 157, 158, 314

Good, Dr. William 121, 125

Granger, Alain 221, 222

Gray, David 98, 120, 122, 124, 215

Greylock 223

Grothaus, Eric 380

Gruen Watch Company 12

Grundig 149, 152, 217, 244

Gwinner, Robert W. 15

**H**

*The Handbook of Plastic Optics* 53, 64, 243

Hengst, Wolfgang 257

Herbol, Charles 16, 23, 46

Hewlett-Packard 59, 60, 80, 87, 89, 91, 94, 95, 283, 303, 308, 317, 327

Hinchman, David F. 68, 70, 71, 92, 119, 120, 132, 159, 196, 212, 222, 236, 240, 247, 256, 272, 284, 285, 290, 302, 309, 321, 330, 331, 333, 343, 344, 346, 347, 350, 353, 361, 380

Hirabayashi, Tami 169

Hitachi 170, 171, 173, 174, 177, 194, 206, 207, 208, 217, 222, 267, 283, 292, 301, 307, 310, 329

Hoff, Arthur M (Stretch) 159, 169, 283, 285, 303, 309, 341, 343, 357, 380

Hoffer, Clarice 343, 380

Hoffmann, Dr. Claus 257, 261, 263

Hohberg, Dr. 151, 152

Honeywell 19, 142, 143, 159, 198, 224, 240, 283, 376

Hood, Terry 119

Horiuchi, Hiroshi 173

Houghton, James R. 319, 329, 339

Howe, R. Edwin 380

Huber, Bruce 204

Hughes, Margaret 24

Hughes, Milford 25, 29

Hummel, Don 136, 138

**I**

IBM 8, 19, 43, 159, 200, 308

Ibuka, Masaru 173, 174

Inoue, Takuji 173

Intel 61

International Molders Union 196, 335

Isaac, Florence 380

Ishii, Fred 305, 306

Itoh, Norio 167, 172, 178

**J**

Jennings, Randy 301

Jones, Philip 169, 170, 171, 173, 174, 175, 176

Jones, Reginald 129

Joseph Schneider Company 132

JVC 170, 171, 173

# K

# L

# M

# R

# S

Wright, William 231, 232, 233, 235, 236

**X**

Xerox 19, 27, 34, 103

**Y**

Yaeger, John 11

Yamaguchi, Y. "Super" 338, 339

**Z**

Zenith 151, 156, 180, 194, 204, 205, 206, 208, 217, 267, 310, 315, 323, 339